T0271355

A Course on Set Theory

Set theory is the mathematics of infinity and part of the core curriculum for mathematics majors. This book blends theory and connections with other parts of mathematics so that readers can understand the place of set theory within the wider context. Beginning with the theoretical fundamentals, the author proceeds to illustrate applications to topology, analysis and combinatorics, as well as to pure set theory. Concepts such as Boolean algebras, trees, games, dense linear orderings, ideals, filters and club and stationary sets are also developed.

Pitched specifically at undergraduate students, the approach is neither esoteric nor encyclopedic. The author, an experienced instructor, includes motivating examples and over 100 exercises designed for homework assignments, reviews and exams. It is appropriate for undergraduates as a course textbook or for self-study. Graduate students and researchers will also find it useful as a refresher or to solidify their understanding of basic set theory.

ERNEST SCHIMMERLING is a Professor of Mathematical Sciences at Carnegie Mellon University, Pennsylvania.

A Course on Set Theory

ERNEST SCHIMMERLING
Carnegie Mellon University, Pennsylvania

CAMBRIDGE
UNIVERSITY PRESS

Shaftesbury Road, Cambridge CB2 8EA, United Kingdom

One Liberty Plaza, 20th Floor, New York, NY 10006, USA

477 Williamstown Road, Port Melbourne, VIC 3207, Australia

314–321, 3rd Floor, Plot 3, Splendor Forum, Jasola District Centre, New Delhi – 110025, India

103 Penang Road, #05–06/07, Visioncrest Commercial, Singapore 238467

Cambridge University Press is part of Cambridge University Press & Assessment, a department of the University of Cambridge.

We share the University's mission to contribute to society through the pursuit of education, learning and research at the highest international levels of excellence.

www.cambridge.org
Information on this title: www.cambridge.org/9781107008175

First published 2011

A catalogue record for this publication is available from the British Library

ISBN   978-1-107-00817-5   Hardback
ISBN   978-1-107-40048-1   Paperback

Contents

vi   *Contents*

Note to the instructor

This book was written for an undergraduate set theory course, which is taught at Carnegie Mellon University every spring. It is aimed at serious students who have taken at least one proof-based mathematics course in any area. Most are mathematics or computer science majors, or both, but life and physical science, engineering, economics and philosophy students have also done well in the course. Other students have used this book to learn the material on their own or as a refresher. Mastering this book and learning a bit of mathematical logic, which is not included, would prepare the student for a first-year graduate level set theory course in the future. The book also contains the minimum amount of set theory that everyone planning to go on in math should know.

I have included slightly more than the maximum amount of material that I have covered in a fifteen-week semester. But I do not reach the maximum every time; in fact, only once. For a slower pace or shorter academic term, one of several options would be to skip Sections 5.6 and 7.2, which are more advanced.

There are over one hundred exercises, more than enough for eight homework assignments, two midterm exams, a final exam and review problems before each exam. Exercises are located at the ends of Chapters 1, 2, 3, 4 and 6. They are also dispersed throughout Chapters 5 and 7. This slight lack of uniformity is tied to the presentation and ultimately makes sense.

In roughly the first half of the book, through Chapter 4, I develop ordinal and cardinal arithmetic starting from the axioms of Zermelo–Fraenkel Set Theory with the Axiom of Choice (ZFC). In other words, this is not a book on what some call *naive set theory*. There is one minor way in which the presentation is not entirely rigorous. Namely, in listing the axioms of ZFC, I use the imprecise word *property* instead of the formal expression *first-order formula* because mathematical logic is not a prerequisite for the course.

Some other textbooks develop the theory of cardinality for as long as possible without using the Axiom of Choice (AC). I do not take this approach because it adds technicalities, which are not used later in the course, and gives students the misleading impression that AC is controversial. By assuming AC from the start, I am able to streamline the theory of cardinality. I may note how AC has been used in a proof but I do not belabor the point. Once, when an alternate proof without AC exists, it is outlined in an exercise.

The second half of the book is designed to give students a sense of the place of set theory within mathematics. Where I draw connections to other fields, I include all the necessary background material. Some of the other areas that come up in Chapter 5 are topology, metric spaces, trees, games and Ramsey theory. The real numbers are constructed using Dedekind cuts in Chapter 6. Chapter 7 introduces the student to filters and ideals, and takes up the combinatorics of uncountable sets. There is no section specifically on Boolean algebra but it is one of the recurring themes in the exercises throughout the book. For the reader's convenience, I have briefly summarized the results on Boolean algebra in the Appendix. All of this material is self-contained.

As I mentioned, before starting this book, students should have at least one semester's worth of experience reading and writing proofs in any area of mathematics; it does not matter which area. They should be comfortable with sets, relations and functions, having seen and used them at a basic level earlier. They should know the difference between integers, rational numbers and real numbers, even if they have not seen them explicitly constructed. And they should have experience with recursive definitions along the integers and proofs by induction on the integers. These notions come up again here but in more sophisticated ways than in a first theoretical mathematics course. There are no other prerequisites. However, because of the emphasis on connections to other fields, students who have taken courses on logic, analysis, algebra, or discrete mathematics will enjoy seeing how set theory and these other subjects fit together. The unifying perspective of set theory will give students significant advantages in their future mathematics courses.

Acknowledgements

As an undergraduate, I studied from *Elements of set theory* by Herbert Enderton and *Set theory: an introduction to independence proofs* by Kenneth Kunen. When I started teaching undergraduate set theory, I recommended *Introduction to set theory* by Karel Hrbacek and Thomas Jech to my students. The reader who knows these other textbooks will be aware of their positive influence.

This book began as a series of handouts for undergraduate students at Carnegie Mellon University. Over the years, they found typographical errors and indicated what needed more explanation, for which I am grateful. I also thank Michael Klipper for proofreading a draft of the book in Spring 2008, when he was a graduate student in the CMU Doctor of Philosophy program.

During the writing of this book, I was partially supported by National Science Foundation Grant DMS-0700047.

1

Preliminaries

In one sense, set theory is the study of mathematics using the tools of mathematics. After millennia of doing mathematics, mathematicians started trying to write down the rules of the game. Since mathematics had already fanned out into many subareas, each with its own terminology and concerns, the first task was to find a reasonable common language. It turns out that everything mathematicians do can be reduced to statements about sets, equality and membership. These three concepts are so fundamental that we cannot define them; we can only describe them. About equality alone, there is little to say other than "two things are equal if and only if they are the same thing." Describing sets and membership has been trickier. After several decades and some false starts, mathematicians came up with a system of laws that reflected their intuition about sets, equality and membership, at least the intuition that they had built up so far. Most importantly, all of the theorems of mathematics that were known at the time could be derived from just these laws. In this context, it is common to refer to laws as *axioms*, and to this particular system as *Zermelo–Fraenkel Set Theory with the Axiom of Choice*, or *ZFC*. In the first unit of the course, through Chapter 4, we examine this system and get some practice using it to build up the theory of infinite numbers.

In another sense, set theory is a part of mathematics like any other, rich in ideas, techniques and connections to other areas. This perspective is emphasized more than the foundational aspects of set theory throughout the course but especially in the second half, Chapters 5–7. There, our choice of topics within set theory is designed to give the reader an impression of the depth and breadth of the subject and where it fits within the whole of mathematics.

To get started, we review some basic notation and terminology. We expect that the reader is familiar with the following notions but perhaps has not seen them expressed in exactly the same way.

Ordered pairs are used everywhere in mathematics, for example, to refer to points on the plane in geometry. The precise meaning of (x, y) is left to the imagination in most other courses but we need to be more specific.

Definition 1.1 $(x, y) = \{\{x\}, \{x, y\}\}$ is the *ordered pair with first coordinate x and second coordinate y.*

It is convenient that (x, y) is defined in terms of sets. After all, this is set theory, so everything should be a set! The main point of the definition is that from looking at $\{\{x\}, \{x, y\}\}$ we can tell which is the first coordinate and which is the second coordinate. Namely, if $\{\{x\}, \{x, y\}\}$ has exactly two elements, then the first coordinate is

$$x = \text{the unique } z \text{ such that } \{z\} \in \{\{x\}, \{x, y\}\}$$

and the second coordinate is

$$y = \text{the unique } z \neq x \text{ such that } \{x, z\} \in \{\{x\}, \{x, y\}\}.$$

And, if $\{\{x\}, \{x, y\}\}$ has just one element, which can only happen if $x = y$, then the first and second coordinates are both

$$x = \text{the unique } z \text{ such that } \{z\} \in \{\{x\}\}.$$

To understand this formula, keep in mind that

$$\{x, y\} = \{y, x\}$$

and

$$\{x, x\} = \{x\}.$$

In particular,

$$\{\{x\}, \{x, x\}\} = \{\{x\}, \{x\}\} = \{\{x\}\}$$

and $\{x\}$ is the only element of $\{\{x\}\}$.

Definition 1.2 $A \times B = \{(x, y) \mid x \in A \text{ and } y \in B\}$ is the *Cartesian product of A and B.*

Definition 1.3 R is a *relation from A to B* iff R is a subset of $A \times B$, that is

$$R \subseteq A \times B.$$

Sometimes, if we know that R is a relation, then we write xRy instead of $(x, y) \in R$. For example, we write

$$\sqrt{2} < \pi$$

not

$$(\sqrt{2}, \pi) \in <$$

because the latter is confusing.

Definition 1.4   Let R be a relation from A to B and $S \subseteq A$.

1. The *domain of R* is

$$\mathrm{dom}(R) = \{x \in A \mid \text{there exists } y \text{ such that } xRy\}.$$

2. The *image of S under R* is

$$R[S] = \{y \in B \mid \text{there exists } x \in S \text{ such that } xRy\}.$$

3. The *range of R* is

$$\mathrm{ran}(R) = \{y \in B \mid \text{there exists } x \text{ such that } xRy\}.$$

Notice that $\mathrm{ran}(R) = R[\mathrm{dom}(R)]$.

Definition 1.5 f is a *function from A to B* iff f is a relation from A to B and, for every $x \in A$, there exists a unique y such that $(x, y) \in f$.

If we happen to know that f is a function, then we write

$$f(x) = y$$

instead of $(x, y) \in f$. When we write $f : A \to B$, it is implicit that f is a function from A to B. In certain situations, we refer to a function f by writing $x \mapsto f(x)$ or $\langle f(x) \mid x \in A \rangle$. There are times when we write f_x instead of $f(x)$; this is when we are thinking of elements x of A as *indices* and $\langle f_x \mid x \in A \rangle$ as an *indexed family*. If the domain of f consists of ordered pairs, then it is common to write $f(x, x')$ instead of $f((x, x'))$. Functions are also called *operations* and *maps*. Some people distinguish between a function $f : A \to B$ and its graph,

$$\text{graph}(f) = \{(x, f(x)) \mid x \in A\},$$

but we do not. To us they are the same, that is, $f = \text{graph}(f)$, as we see from Definition 1.5.

Definition 1.6 If $f : A \to B$ is a function and $S \subseteq A$, then the *restriction of f to S* is

$$f \upharpoonright S = \{(x, f(x)) \mid x \in S\}.$$

Definition 1.7 Let $f : A \to B$ be a function.

1. f is an *injection* iff for all $x, x' \in A$, if $x \neq x'$, then $f(x) \neq f(x')$.
2. f is a *surjection* iff for every $y \in B$, there exists $x \in A$ such that $f(x) = y$.
3. f is a *bijection* iff f is both an injection and a surjection.

Injections are also called *one-to-one* functions. Surjections from A to B are also called functions from A *onto* B. Bijections are also called *one-to-one correspondences*.

Definition 1.8 If f is an injection from A to B, then we write f^{-1} for the unique injection $g : f[A] \to A$ with the property that $g(f(x)) = x$ for every $x \in A$. In other words,

$$f^{-1} = \{(f(x), x) \mid x \in A\}.$$

Finally, we assume that the reader has good intuition about the set of integers,

$$\mathbb{Z} = \{\ldots, -2, -1, 0, 1, 2, \ldots\},$$

the set of rational numbers,

$$\mathbb{Q} = \{m/n \mid m, n \in \mathbb{Z} \text{ and } n \neq 0\}$$

and the set of real numbers, $\mathbb{R}$. One thing we will do in this course is define all these kinds of numbers, starting from the natural numbers 0, 1, 2, 3, 4, etc. Each natural number will be the set of natural numbers that precedes it. Thus $0 = \emptyset$, where $\emptyset$ is the set with no members. After that, $1 = \{0\}$, $2 = \{0, 1\}$, $3 = \{0, 1, 2\}$,

$4 = \{0, 1, 2, 3\}$, etc. This happens to be very convenient because then

$$m < n \iff m \in n.$$

In other words, the usual ordering on the natural numbers coincides with membership.

We use natural numbers to denote cardinality, for example, when we say, "Lance Armstrong won the Tour de France seven times." And we use natural numbers to denote order, for example, when we say, "the attorney general is seventh in the presidential line of succession." Another thing we will do in this course is extend the notions of cardinality and order into the infinite. Finite cardinal and ordinal numbers are basically the same thing; one could say that the difference between "seven" and "seventh" is just grammatical. However, the difference between infinite cardinal and ordinal numbers is more profound, as we will explain in Chapters 3 and 4.

Exercises

Exercise 1.1 If R is a relation, then we define

$$R^{-1} = \{(y, x) \mid xRy\}.$$

Give an example where R is a function but R^{-1} is not.

Exercise 1.2 How many functions whose domain is the empty set are there? In other words, given a set B, how many functions $f : \emptyset \to B$ are there?

Exercise 1.3 Explain why $(x, y, z) = (x, (y, z))$ is a reasonable definition of an *ordered triple*.

Exercise 1.4 Equivalence relations play an important role in this book. We assume that the reader has studied them before but this exercise reviews all the necessary definitions and facts. Let A be a set and R be a *relation on A*, that is, $R \subseteq A \times A$. Then:

- R is a *reflexive relation on A* iff for every $x \in A$, xRx.
- R is a *symmetric relation on A* iff for all $x, y \in A$, if xRy, then yRx.

- R is a *transitive relation on A* iff for all $x, y, z \in A$, if xRy and yRz, then xRz.[1]
- R is an *equivalence relation on A* iff R is a reflexive, symmetric and transitive relation on A.

Assuming that R is an equivalence relation on A, for every $x \in A$, we define the *equivalence class of x* to be

$$[x]_R = \{y \in A \mid xRy\}.$$

It is also standard to write

$$A/R = \{[x]_R \mid x \in A\}.$$

A *partition of A* is a family $\mathcal{F}$ of non-empty subsets of A such that

- A is the union of $\mathcal{F}$, that is,

$$A = \bigcup \mathcal{F} = \{x \mid \text{there exists } X \in \mathcal{F} \text{ such that } x \in X\}$$

and
- the elements of $\mathcal{F}$ are pairwise disjoint, that is, for all $X, Y \in \mathcal{F}$, if $X \neq Y$, then $X \cap Y = \emptyset$.

Now here are the exercises:

1. Let R be an equivalence relation on A. Prove that A/R is a partition of A.
2. Let $\mathcal{F}$ be a partition of A. Prove that there exists a unique equivalence relation R such that $\mathcal{F} = A/R$.

---

[1] Later in the book we will define *transitive set*, which is different from *transitive relation*. Unfortunately, it will be important to pay attention to this subtle difference in terminology.

2
ZFC

In the most general terms, when we talk about a mathematical *theory*, we have in mind a collection of *axioms* in a certain *language*. The *language of set theory* has two symbols, $=$ and $\in$, although sometimes we add symbols that are defined in terms of these two to make things easier to read. For example, we write $A \subseteq B$ when we mean that, for every x, if $x \in A$, then $x \in B$.

Zermelo–Fraenkel Set Theory with the Axiom of Choice, or *ZFC* for short, is a certain theory in the language of set theory that we will describe in this chapter. There are infinitely many axioms of ZFC, each of which says something rather intuitive about sets, equality and membership. In our list below, some axioms of ZFC are presented individually whereas others are presented as *schemes* for generating infinitely many axioms. One last comment about terminology before we begin: throughout the course,

$$\text{set} = \text{collection} = \text{family}$$

and

$$\text{member} = \text{element}.$$

Also, the three phrases,

- x belongs to A,
- x is an element of A and
- x is a member of A,

all mean the same thing, namely $x \in A$.

Empty Set Axiom

This axiom says that there is a unique set without members. Formally, it is written

$$\exists! A \; \forall x \; (x \notin A).$$

In plain English, this says:

There exists a unique A such that, for every x,
x is not an element of A.

The unique set without elements is written $\emptyset$.

Extensionality Axiom

This axiom says that two sets are equal if they have the same members. Formally, it is written

$$\forall A \; \forall B \; [\; \forall x \; (x \in A \iff x \in B) \implies A = B \;].$$

Because we defined

$$A \subseteq B \iff \forall x \; (x \in A \implies x \in B),$$

another way to write the Extensionality Axiom is

$$\forall A \; \forall B \; [\; (A \subseteq B \text{ and } B \subseteq A) \implies A = B \;].$$

In other words, two sets are equal if each is a subset of the other.

By logic alone, if $A = B$, then A and B have the same members. Combining this fact with the Extensionality Axiom, we have that

$$\forall A \; \forall B \; [\; \forall x \; (x \in A \iff x \in B) \iff A = B \;].$$

Equivalently,

$$\forall A \; \forall B \; [\; (A \subseteq B \text{ and } B \subseteq A) \iff A = B \;].$$

Pairing Axiom

This axiom allows us to form singletons and unordered pairs. Its formal statement is

$$\forall x \; \forall y \; \exists! A \; \forall z \; [z \in A \iff (z = x \text{ or } z = y)].$$

If $x \neq y$, then we write $\{x, y\}$ for the unique set whose only members are x and y and call it an *unordered pair*. We always write $\{x\}$ instead of $\{x, x\}$ and call it a *singleton*. At this point, it makes sense to define the first three natural numbers $0 = \emptyset$, $1 = \{0\}$ and $2 = \{0, 1\}$. We can also justify defining *ordered pairs* by setting

$$(x, y) = \{\{x\}, \{x, y\}\}$$

whenever we are given x and y as we did in Definition 1.1. As a reminder, when $x = y$, what we really have is

$$(x, x) = \{\{x\}\}.$$

Notice that, based on this definition, when we write (x, y), we can tell that x is the *first coordinate* and y is the *second coordinate*. Formally, this means we can prove that for all x, y, x' and y',

$$(x, y) = (x', y') \iff (x = x' \text{ and } y = y').$$

Union Axiom

This axiom allows us to form unions. Its formal statement is

$$\forall \mathcal{F} \; \exists! A \; \forall x \; [x \in A \iff \exists Y \in \mathcal{F} \; (x \in Y)].$$

We write $\bigcup \mathcal{F}$ for the unique set whose members are exactly the members of the members of $\mathcal{F}$. In other words,

$$\bigcup \mathcal{F} = \{x \mid \text{there exists } Y \in \mathcal{F} \text{ such that } x \in Y\}.$$

It is important to note that, in the Union Axiom, the family $\mathcal{F}$ is allowed to be infinite. We often use different notation when $\mathcal{F}$ is finite. For example, we define

$$A \cup B = \bigcup \{A, B\}$$

and

$$A \cup B \cup C = \bigcup \{A, B, C\}.$$

At this point, we can define the remaining natural numbers

$$3 = 2 \cup \{2\} = \{0, 1, 2\},$$

$$4 = 3 \cup \{3\} = \{0, 1, 2, 3\},$$

$$5 = 4 \cup \{4\} = \{0, 1, 2, 3, 4\}$$

and, in general,

$$n + 1 = n \cup \{n\} = \{0, \dots, n\}.$$

Power Set Axiom

This axiom allows us to form the set of all subsets of a given set. Its formal statement is

$$\forall A\ \exists! \mathcal{F}\ \forall X\ (X \in \mathcal{F} \iff X \subseteq A).$$

We write $\mathcal{P}(A)$ for the unique set of subsets of A. In other words,

$$\mathcal{P}(A) = \{X \mid X \subseteq A\}.$$

We call $\mathcal{P}(A)$ the *power set* of A. As an example, let us see what happens when we start with the empty set and take power sets over and over. Define

$$V_0 = \emptyset,$$

$$V_1 = \mathcal{P}(V_0) = \{\emptyset\},$$

$$V_2 = \mathcal{P}(V_1) = \{\emptyset, \{\emptyset\}\},$$

$$V_3 = \mathcal{P}(V_2) = \{\emptyset, \{\emptyset\}, \{\{\emptyset\}\}, \{\emptyset, \{\emptyset\}\}\}$$

and, in general,

$$V_{n+1} = \mathcal{P}(V_n).$$

The sets V_n will come up again later.

Comprehension Scheme

This axiom scheme gives us a way to form specific subsets of a given set. It says the following.

For each "property" $P(x)$, the following is an axiom:

$$\forall A\ \exists! B\ \forall x\ [x \in B \iff (x \in A \text{ and } P(x))].$$

Notice that the word "property" appears in quotes. There are infinitely many properties, which is why ZFC has infinitely many axioms. We will not give a formal definition of "property" because it involves first-order logic, which is not a prerequisite. It is enough for students in this course to depend on their intuition about the meaning of "property". Given a property $P(x)$, we write

$$\{x \in A \mid P(x)\}$$

for the set of elements x of A for which $P(x)$ is true. For example,

$$\{x \in 10 \mid x \text{ is even}\} = \{0, 2, 4, 6, 8\}$$

and

$$\{x \in 10 \mid x \text{ is odd}\} = \{1, 3, 5, 7, 9\}.$$

It is important to note that the Comprehension Scheme does not, in general, permit us to define sets by writing $\{x \mid P(x)\}$. In fact, for some $P(x)$, what we think of as $\{x \mid P(x)\}$ is not a set. For example $\{x \mid x = x\}$ is not a set by the result in Exercise 2.6.

At this point, we are justified in making many familiar definitions. For example, Definition 1.2 says that, given sets A and B, we have the Cartesian product

$$A \times B = \{(x, y) \mid x \in A \text{ and } y \in B\}.$$

In order to see that this is a legitimate definition, note that if $x \in A$ and $y \in B$, then

$$(x, y) = \{\{x\}, \{x, y\}\} \in \mathcal{P}\left(\mathcal{P}\left(A \cup B\right)\right).$$

Thus $A \times B =$

$$\{z \in \mathcal{P}\left(\mathcal{P}\left(A \cup B\right)\right) \mid z = (x, y) \text{ for some } x \in A \text{ and } y \in B\},$$

which means the definition of $A \times B$ is justified by a combination of the axioms we have listed so far. See Exercise 2.4.

Infinity Axiom

In an indirect way, this axiom tells us that the set of natural numbers exists. Its formal statement is

$$\exists I \left[\emptyset \in I \text{ and } \forall x \left(x \in I \implies x \cup \{x\} \in I\right) \right].$$

We say that a set I is *inductive* if I is a witness to the Infinity Axiom. In other words,

$$I \text{ is inductive} \iff \left[\emptyset \in I \text{ and } \forall x \left(x \in I \implies x \cup \{x\} \in I\right) \right].$$

Proposition 2.1   *There is a unique inductive set I such that $I \subseteq J$ for every inductive set J.*

The proof of Proposition 2.1 is broken up into smaller steps in Exercise 2.5. The set of natural numbers is defined to be the unique inductive set that is a subset of every other inductive set. We write ω (lower case Greek omega) for the set of natural numbers. In other words,

$$\omega = \{0, 1, 2, 3, \dots\}.$$

It is also common for mathematicians to write $\mathbb{N}$ instead of ω although we will not. Not only does the Infinity Axiom allow us to define ω, it implies that we can prove statements by induction on $n \in \omega$ and make recursive definitions for $n \in \omega$ just as you have done in your other mathematics courses.

To see the relationship with induction, suppose we wish to prove that a given property $P(n)$ holds for every natural number n. By Proposition 2.1 and the definition of ω, it would be enough to show that $\{n \in \omega \mid P(n)\}$ is an inductive set. In other words, show that $P(0)$ holds and if $P(n)$ holds, then so does $P(n+1)$.

Replacement Scheme

This is a scheme for generating infinitely more axioms.

For each "property" $P(x, y)$, the following is an axiom:

$$\forall A \left[(\forall x \in A \; \exists y \; P(x, y)) \implies (\exists B \; \forall x \in A \; \exists y \in B \; P(x, y)) \right].$$

The same comments we made about meaning of "property" when we discussed the Comprehension Scheme apply here too.

Because we will not emphasize how the Replacement Scheme is used later in the book, let us give a concrete example and some intuition here. Suppose we want to define

$$V_\omega = \bigcup \{V_n \mid n \in \omega\}.$$

That is, we want to let V_ω be the union of the infinite family

$$\{V_n \mid n \in \omega\}.$$

This family is supposed to be the range of the infinite sequence

$$\langle V_n \mid n \in \omega \rangle.$$

But why does this infinite sequence exist? In other words, why is it a set? Given a particular natural number, say 5, we can prove from the other axioms that the finite sequence

$$\langle V_m \mid m < 5 \rangle = \langle V_0, V_1, V_2, V_3, V_4 \rangle$$

exists. The Replacement Scheme can be used to make the leap from infinitely many finite sequences to one infinite sequence. To see what we mean, let $P(x, y)$ be the following property: *x is a natural number and there exists a function f with domain*

$$\mathrm{dom}(f) = x + 1 = \{0, \ldots, x\}$$

such that

- $f(0) = \emptyset$,
- *for every* $n < x$, $f(n+1)$ *is the power set of* $f(n)$, *and*
- $f(x) = y$.

Then $P(n, V_n)$ holds for every $n \in \omega$. By the Replacement Scheme, there is a set B such that, for every $n \in \omega$, $V_n \in B$. Now use the Comprehension Scheme to define

$$\langle V_n \mid n < \omega \rangle = \{(x, y) \in \omega \times B \mid P(x, y)\}.$$

Finally, use the Union Axiom to define V_ω as we originally wanted to do.

A slogan that captures the intuition behind the Replacement Scheme is: *If an assignment looks like a function and its domain is a set, then its range is also a set, so it really is a function.* In the example above, we used the Replacement Scheme to see that the assignment $n \mapsto V_n$ really is a function with domain ω and range $\{V_n \mid n \in \omega\}$.

Foundation Axiom

This axiom says that if S is a non-empty set, then there exists $x \in S$ such that, for every $y \in S$, $y \notin x$. In symbols,

$$\forall S \ (S \neq \emptyset \implies \exists x \in S \ \forall y \in S \ (y \notin x)).$$

For example, should $S = 2$, the only witness to the Foundation Axiom would be $x = 0$ because $2 = \{0, 1\}$ and $0 \in \{0\} = 1$ but $1 \notin \emptyset = 0$. On the other hand, when $S = \{0, \{1\}\}$, both elements of S satisfy the requirement of the Foundation Axiom because $0 \notin \{1\}$ and $\{1\} \notin 0$.

The Foundation Axiom turns out to be equivalent to the statement that there is no sequence $\langle x_n \mid n \in \omega \rangle$ such that

$$\cdots x_{n+1} \in x_n \in \cdots \in x_1 \in x_0.$$

In particular, it implies that no set is an element of itself. For otherwise, if $x \in x$, then

$$\cdots x \in x \in x \in x,$$

which contradicts the Foundation Axiom. We will make additional comments about this axiom later but it is not a focus of the book.

Axiom of Choice

This axiom says that every family of sets has a choice function.

$$\forall \mathcal{F} \ \exists \text{ function } c \ \forall A \in \mathcal{F} \ (A \neq \emptyset \implies c(A) \in A).$$

The function c is called a *choice function* for $\mathcal{F}$ because it chooses an element $c(A)$ out of every non-empty A that belongs to $\mathcal{F}$. From experience, given finitely many non-empty sets, you can pick one element from each. The Axiom of Choice says this is possible even if you start with infinitely many non-empty sets.

The first time the Axiom of Choice is used is in Chapter 4. There, it is essential for being able to assign a numerical cardinality (size) to each set, which is one of the most important things we do in this book. As we go along, we will point out where and how the Axiom of Choice is used.

Exercises

In the following exercises, you only need to be attentive to the axioms of ZFC when so instructed, namely, in Exercises 2.1, 2.4, 2.6, the second part of 2.8 and the first part of 2.11. Otherwise, use the same style of mathematical argumentation as in your other proof-based mathematics courses without reference to ZFC.

Exercise 2.1 Prove that the following theories are equivalent.

1. Empty Set Axiom + Extensionality Axiom.
2. $\exists A \, \forall x \, (x \notin A)$ + Extensionality Axiom.

Hint: Obviously 1 implies 2. In the other direction, uniqueness is the issue.

Exercise 2.2 Recall that $0 = \emptyset$ and $n+1 = \{0, \dots, n\}$ for every $n \in \omega$. Recall also that $V_0 = \emptyset$ and $V_{n+1} = \mathcal{P}(V_n)$ for every $n \in \omega$.

1. Prove by induction that if $n \in \omega$, then $\mathcal{P}(n)$ has 2^n elements.
2. List the elements of V_4.
3. Make a conjecture regarding the size of V_n. Then prove your conjecture by induction on $n \in \omega$.

Exercise 2.3 Define z to be a *transitive set* iff for all x and y, if $x \in y$ and $y \in z$, then $x \in z$. This is equivalent to saying that, for every y, if $y \in z$, then $y \subseteq z$.

1. Prove that V_n is a transitive set for every $n \in \omega$.
2. Prove that $V_n \subseteq V_{n+1}$ for every $n \in \omega$.
3. Prove that $V_n \cap \omega = n$ for every $n \in \omega$.

Exercise 2.4 Give a detailed explanation of how the definitions of $A \times B$, $\mathrm{dom}(R)$ and $R[S]$ given in Definitions 1.2 and 1.4 are justified by the axioms of ZFC.

Exercise 2.5 Prove Proposition 2.1 by showing the following.

1. If I and J are inductive sets, then $I \cap J$ is an inductive set.
2. Suppose that K is an inductive set. Let

$$\mathcal{F} = \{J \in \mathcal{P}(K) \mid J \text{ is an inductive set}\}$$

and

$$I = \{x \in K \mid \forall J \in \mathcal{F} \; (\, x \in J \,)\}.$$

Prove the following statements.

(a) I is an inductive set.
(b) For every J, if J is an inductive set, then $I \subseteq J$.
(c) Suppose that I' is an inductive set and, for every inductive set J,

$$I' \subseteq J.$$

Then $I' = I$.

Exercise 2.6 Prove that there is no set V such that $x \in V$ for every set x. (This shows that $\{x \mid x = x\}$ is not a set.) If you use the Foundation Axiom in your proof, then find a second proof that does not use the Foundation Axiom.

Exercise 2.7 In general, define the intersection of a non-empty family $\mathcal{G}$ to be

$$\bigcap \mathcal{G} = \{a \mid \forall X \in \mathcal{G} \ (a \in X)\}.$$

Let S be a set and $\mathcal{F} \subseteq \mathcal{P}(S)$. Assume that $\mathcal{F} \neq \emptyset$. Prove that

$$S - \bigcap \mathcal{F} = \bigcup\{S - X \mid X \in \mathcal{F}\}$$

and

$$S - \bigcup \mathcal{F} = \bigcap\{S - X \mid X \in \mathcal{F}\}.$$

Exercise 2.8 In general, define

$$^A B = \{f \mid f \text{ is a function from } A \text{ to } B\}.$$

Consider a function of the form $(a, b) \mapsto S_{(a,b)}$ with domain $A \times B$. In other words, consider an indexed family

$$\langle S_{(a,b)} \mid (a, b) \in A \times B \rangle.$$

1. Prove that

$$\bigcap_{a \in A} \bigcup_{b \in B} S_{(a,b)} = \bigcup_{f \in {}^A B} \bigcap_{a \in A} S_{(a,f(a))}.$$

2. Assume in addition that

$$S_{(a,b)} \cap S_{(a,b')} = \emptyset$$

whenever $a \in A$ and $b, b' \in B$ but $b \neq b'$. Now prove the same equation as in part 1 but without using the Axiom of Choice.

Exercise 2.9 By definition, the *symmetric difference* of A and B is

$$A \triangle B = (A - B) \cup (B - A).$$

Verify the distributive law

$$A \cap (B \triangle C) = (A \cap B) \triangle (A \cap C).$$

Remark: This is one step in showing that if S is a set, then $\mathcal{P}(S)$

is a ring if addition is taken to be symmetric difference and multiplication is taken to be intersection. We only mention this for readers who happen to know some abstract algebra.

Exercise 2.10 Define a relation E on $\mathcal{P}(\omega)$ according to the formula

$$(x, y) \in E \iff x \vartriangle y \text{ is finite.}$$

(See Exercise 2.9 for the definition of $x \vartriangle y$.)

1. Prove that E is an equivalence relation on $\mathcal{P}(\omega)$.

2. We write $[x]_E$ for the equivalence class of x, in other words,

$$[x]_E = \{y \in \mathcal{P}(\omega) \mid xEy\}.$$

Prove that for every E-equivalence class $[x]_E$ is infinite.

3. By $\mathcal{P}(\omega)/E$ we mean the family of E-equivalence classes, that is,

$$\mathcal{P}(\omega)/E = \{[x]_E \mid x \in \mathcal{P}(\omega)\}.$$

Prove that $\mathcal{P}(\omega)/E$ is infinite.

Exercise 2.11 Let A be a set. By recursion on $n \in \omega$, define

$$B_0 = A$$

and

$$B_{n+1} = \bigcup B_n.$$

Let

$$C = \bigcup \{B_n \mid n < \omega\}.$$

1. Which axioms of ZFC are used to see that C is a set?
2. Prove that C is a transitive set. (See Exercise 2.3 for the definition of *transitive set*.)
3. Prove that if D is a transitive set and $A \subseteq D$, then $C \subseteq D$.

We call C the *transitive closure* of A.

Beginning exercises on Boolean algebra

To understand Exercises 2.12 and 2.13 below, and many exercises
later in the book,[1] you must first read the following definitions
and example.

A *Boolean algebra* is a 6-tuple of the form

$$\mathbb{B} = (B, \vee, \wedge, \neg, \bot, \top)$$

where B is a set, $\vee$ and $\wedge$ are binary operations on B, $\neg$ is a unary
operation on B, $\bot$ and $\top$ are distinct elements of B, and for all
$X, Y, Z \in B$, the following ten laws hold.

Associativity

$$X \vee (Y \vee Z) = (X \vee Y) \vee Z$$
$$X \wedge (Y \wedge Z) = (X \wedge Y) \wedge Z$$

Commutativity

$$X \vee Y = Y \vee X$$
$$X \wedge Y = Y \wedge X$$

Distributivity

$$X \vee (Y \wedge Z) = (X \vee Y) \wedge (X \vee Z)$$
$$X \wedge (Y \vee Z) = (X \wedge Y) \vee (X \wedge Z)$$

Identity

$$X \vee \bot = X$$
$$X \wedge \top = X$$

Complementation

$$X \vee \neg X = \top$$
$$X \wedge \neg X = \bot$$

Each Boolean algebra, $\mathbb{B} = (B, \vee, \wedge, \neg, \bot, \top)$, has an associated
Boolean algebra relation, $\preccurlyeq$, which is defined by

$$X \preccurlyeq Y \iff X = X \wedge Y$$

for all $X, Y \in B$. Here is some special terminology for Boolean
algebras that will be used in the exercises. If $A \in B$, then we say

---

[1] The Appendix lists the exercises to which we are referring.

that A is an *atom* iff $A \neq \bot$ and, for every $X \in B$, if $X \preccurlyeq A$, then $X = \bot$ or $X = A$. We say that $\mathbb{B}$ is *finite* iff B is finite.

Given Boolean algebras

$$\mathbb{B} = (B, \vee_{\mathbb{B}}, \wedge_{\mathbb{B}}, \neg_{\mathbb{B}}, \bot_{\mathbb{B}}, \top_{\mathbb{B}})$$

and

$$\mathbb{C} = (C, \vee_{\mathbb{C}}, \wedge_{\mathbb{C}}, \neg_{\mathbb{C}}, \bot_{\mathbb{C}}, \top_{\mathbb{C}}),$$

an *isomorphism from* $\mathbb{B}$ *to* $\mathbb{C}$ is defined to be a function f such that

- f is a bijection from B to C and
- for all $X, Y \in B$,

$$f(X \vee_{\mathbb{B}} Y) = f(X) \vee_{\mathbb{C}} f(Y),$$

$$f(X \wedge_{\mathbb{B}} Y) = f(X) \wedge_{\mathbb{C}} f(Y),$$

$$f(\neg_{\mathbb{B}} X) = \neg_{\mathbb{C}} f(X),$$

$$f(\bot_{\mathbb{B}}) = \bot_{\mathbb{C}}$$

and

$$f(\top_{\mathbb{B}}) = \top_{\mathbb{C}}.$$

Example You should work out the details of the following assertions before attempting the exercises that follow. Let $S \neq \emptyset$ and put

$$\mathcal{B}(S) = (\mathcal{P}(S), \cup, \cap, -, \emptyset, S)$$

where $-X$ means $S - X$ in this context. Then $\mathcal{B}(S)$ is a Boolean algebra. It is called the *Boolean algebra of subsets of* S. If S is finite and has n elements, then $\mathcal{P}(S)$ has 2^n elements hence $\mathcal{B}(S)$ is finite. For $\mathcal{B}(S)$, the Boolean algebra relation amounts to

$$X \preccurlyeq Y \iff X = X \cap Y \iff X \subseteq Y.$$

The atoms of $\mathcal{B}(S)$ are exactly the singletons $\{a\}$ for $a \in S$. The function $X \mapsto S - X$ is an isomorphism from $\mathcal{B}(S)$ to the Boolean algebra

$$(\mathcal{P}(S), \cap, \cup, -, S, \emptyset).$$

Notice that union and intersection are exchanged, as are $\emptyset$ and S.

Exercise 2.12 Consider an arbitrary finite Boolean algebra

$$\mathbb{B} = (B, \vee, \wedge, \neg, \bot, \top).$$

Let S be the set of atoms of $\mathbb{B}$.

1. Prove that if $X \in B$ and $X \neq \bot$, then there exists $A \in S$ such that $A \preccurlyeq X$.

2. Let $X \in B$. Suppose that $X \neq \bot$ and

$$\{A \in S \mid A \preccurlyeq X\} = \{A_1, \ldots, A_m\}.$$

Let

$$Y = A_1 \vee \cdots \vee A_m.$$

Prove

$$X = Y.$$

Hint: Certain basic facts about sets generalize to Boolean algebras. For example, for sets we know that

$$\text{if } X \subseteq Y \text{ and } Y \subseteq X, \text{ then } X = Y,$$

and for Boolean algebras we have that

$$\text{if } X \preccurlyeq Y \text{ and } Y \preccurlyeq X, \text{ then } X = Y.$$

The reason is that if $X \preccurlyeq Y$ and $Y \preccurlyeq X$, then

$$X = X \wedge Y = Y \wedge X = Y$$

by the commutativity law for $\wedge$ and the definition of $\preccurlyeq$. The moral is that you should base your intuition about Boolean algebras on what you already know about Boolean algebras of sets. Of course, ultimately, you need to prove your intuition is correct using just the laws of Boolean algebras. It should also be said that the solution to this exercise is relatively long so organizing your answer into a well-chosen series of lemmas would be very helpful.

3. Let $f : B \to \mathcal{P}(S)$ be defined by

$$f(X) = \{A \in S \mid A \preccurlyeq X\}.$$

Prove that f is an isomorphism from $\mathbb{B}$ to $\mathcal{B}(S)$. *Remark:* This explains why intuition coming from Boolean algebras of sets really is valuable intuition about finite Boolean algebras.

Exercise 2.13   As in Exercise 2.10, let E be the equivalence relation on $\mathcal{P}(\omega)$ defined by

$$x \, E \, y \iff x \, \triangle \, y \text{ is finite.}$$

1. Prove that the following table of equations determines a Boolean algebra $\mathbb{B} = (B, \vee, \wedge, \neg, \bot, \top)$.

$$B = \mathcal{P}(\omega)/E$$
$$[x]_E \vee [y]_E = [x \cup y]_E$$
$$[x]_E \wedge [y]_E = [x \cap y]_E$$
$$\neg [x]_E = [\omega - x]_E$$
$$\bot = [\emptyset]_E$$
$$\top = [\omega]_E$$

Before proving the laws of Boolean algebras, you must show that the operations $\vee$, $\wedge$ and $\neg$ are well-defined by the equations listed above. So part of what you must show is that if $x \, E \, x'$ and $y \, E \, y'$, then

$$(x \cup y) \, E \, (x' \cup y'),$$

$$(x \cap y) \, E \, (x' \cap y')$$

and

$$(\omega - x) \, E \, (\omega - x').$$

Remark: This is an example of a *quotient* Boolean algebra. In the literature, it is referred to as $\mathcal{P}(\omega)/\text{Finite}$.

2. Prove that the Boolean algebra $\mathbb{B}$ that was defined in part 1 has no atoms.

3
Order

At the end of Chapter 1, we gave examples of sentences in English that illustrated the difference between ordinal and cardinal numbers. Let us expand on our example of ordinal numbers, which involved the presidential line of succession:

1st	Vice President
2nd	Speaker of the House
3rd	President pro tempore of the Senate
4th	Secretary of State
5th	Secretary of the Treasury
6th	Secretary of Defense
7th	Attorney General
8th	Secretary of the Interior
9th	Secretary of Agriculture
10th	Secretary of Commerce
11th	Secretary of Labor
12th	Secretary of Health and Human Services
13th	Secretary of Housing and Urban Development
14th	Secretary of Transportation
15th	Secretary of Energy

For fun, here are a few more:

16th	Secretary of Education
17th	Secretary of Veterans Affairs
18th	Secretary of Homeland Security

What is the point? In English, we usually count ordinal numbers 1st, 2nd, 3rd, etc. However, sometimes it makes sense to count 0th, 1st, 2nd, 3rd, etc. For example, at the top of the presidential line of succession, we really have:

0th	President

In set theory, we start counting ordinals from 0. For example, in the sequence of length 6,

$$\langle x_0, x_1, x_2, x_3, x_4, x_5 \rangle = \langle 11, 6, 18, 9, 72, 31 \rangle$$

the 0th number is $x_0 = 11$, the 1st number is $x_1 = 6$, etc., and the 5th and final number is $x_5 = 31$. Keep in mind that, in plain English, it would be very strange to say that the fifth item is the last in a list of six items!

In set theory, we also continue counting ordinal numbers past all the finite ordinal numbers. For example, ω is the first infinite ordinal number, followed by $\omega + 1$, $\omega + 2$, etc. It takes a bit of theoretical work to make concrete sense of this idea, so this is what we do first.

3.1 Wellorderings

Definition 3.1 Let A be a set and $\prec$ be a relation on A.

1. $(A, \prec)$ is *transitive* iff for all $x, y, z \in A$, if $x \prec y$ and $y \prec z$, then $x \prec z$.
2. $(A, \prec)$ is *irreflexive* iff for every $x \in A$, $x \not\prec x$.
3. $(A, \prec)$ is *total* iff for all $x, y \in A$, either $x \prec y$ or $x = y$ or $y \prec x$.

4. $(A, \prec)$ is a *strict linear ordering* iff it is transitive, irreflexive and total.

Definition 3.2 Let $\prec$ and $\preccurlyeq$ be two relations on A. Suppose that

$$x \preccurlyeq y \iff (x \prec y \text{ or } x = y)$$

for all $x, y \in A$. Then $(A, \preccurlyeq)$ is a *linear ordering* iff $(A, \prec)$ is a strict linear ordering.

The definition tells us how to pass from a strict linear ordering to its associated linear ordering. In the other direction, we have the fact that if $(A, \preccurlyeq)$ is a linear ordering and

$$x \prec y \iff (x \preccurlyeq y \text{ and } x \neq y)$$

for all $x, y \in A$, then $(A, \prec)$ is a strict linear ordering and $(A, \preccurlyeq)$ is the linear ordering associated to $(A, \prec)$. Occasionally, we might drop the word *strict* when it is clear from the choice of symbols or context which we mean.

Here is another remark on notation. If we are given that $(A, \prec)$ is a strict linear ordering, then we might write $\preccurlyeq$ without bothering to explain that it is the linear ordering associated to $(A, \prec)$ even though officially we should explain. Suppose, instead, that we are told that (A, R) is a strict linear ordering. It is unlikely that we would write $\underline{R}$ for the associated linear ordering because it looks so strange. If, for some reason, we wrote $\underline{R}$, then we could not assume that the reader knows what we mean, so we would have to explain. Officially, $\prec$ and $\preccurlyeq$ are two completely different symbols even though $\preccurlyeq$ looks like a combination of $\prec$ and $=$.

Definition 3.3 Let $(A, \prec)$ be a strict linear ordering, $S \subseteq A$ and $x \in S$. Then x is the $\prec$-*least element of S* iff for every $y \in S$, $x \preccurlyeq y$.

Definition 3.4 $(A, \prec)$ is a *wellordering* iff it is a strict linear ordering and, for every non-empty $S \subseteq A$, S has a $\prec$-least element.

The extra property that turns a strict linear ordering into a wellordering is called *wellfoundedness*. Every finite linear ordering is a wellordering. Also, the usual linear ordering of the natural numbers is a wellordering. On the other hand, the usual linear ordering on the set of integers,

$$\mathbb{Z} = \{\cdots - 2, -1, 0, 1, 2, \dots\},$$

is obviously not a wellordering because there is no least integer. However, if we define a brand new relation $\prec$ on $\mathbb{Z}$ by

$$0 \prec 1 \prec -1 \prec 2 \prec -2 \prec 3 \prec -3 \prec \cdots,$$

then $(\mathbb{Z}, \prec)$ is a wellordering.

Lemma 3.5 *Let $(A, \prec)$ be a strict linear ordering. Then $(A, \prec)$ is a wellordering iff there is no sequence $\langle x_n \mid n \in \omega \rangle$ such that, for every $n \in \omega$, $x_{n+1} \prec x_n$.*

Proof First suppose that $(A, \prec)$ is a wellordering. Consider an arbitrary sequence $\langle x_n \mid n \in \omega \rangle$. Let $S = \{x_n \mid n \in \omega\}$. Let y be the $\prec$-least element of S. Say $y = x_n$. Then $x_n \preccurlyeq x_{n+1}$. Hence $x_{n+1} \not\prec x_n$.

Second, suppose that $(A, \prec)$ is not a wellordering. Therefore, there exists $S \subseteq A$ such that $S \neq \emptyset$ and S does not have a $\prec$-least element. Define a sequence x_n by recursion on $n \in \omega$ as follows. For the base case, let $x_0 \in S$ if possible. Otherwise, leave x_0 undefined. For the successor case, let $x_{n+1} \in S$ with $x_{n+1} \prec x_n$ if possible. Otherwise, leave x_{n+1} undefined.

We claim that, for every $n \in \omega$, x_n is defined, $x_n \in S$ and, if $n \geq 1$, then $x_n \prec x_{n-1}$. We prove this claim by induction on $n \in \omega$. The base case is $n = 0$. Since $S \neq \emptyset$, it is possible to let $x_0 \in S$. So that is what we did. The successor case has the induction hypothesis that x_n is defined and $x_n \in S$. Since S does not have a $\prec$-least element, it is possible to let $x_{n+1} \in S$ with $x_{n+1} \prec x_n$. So that is what we did. Note that $x_{(n+1)-1} = x_n$. $\square$

In the second part of the previous proof, we made a definition by recursion on $n \in \omega$ and proved a statement by induction on $n \in \omega$. Our ability to do this is tied to the fact that the usual ordering of ω is a wellordering. The following fundamental theorems put the phenomena of induction and recursion in the general context of wellorderings.

Theorem 3.6 (Proofs by induction) *Let $(A, \prec)$ be a wellordering. Let $P(x)$ be a statement about a variable x. Suppose that, for* *every $y \in A$,*

$$(\forall x \prec y \ P(x) \ holds) \implies P(y) \ holds.$$

Then, for every $y \in A$, $P(y)$ holds.

Proof Let $S = \{x \in A \mid P(x) \text{ does not hold}\}$. For contradiction, suppose that $S \neq \emptyset$. Let y be the $\prec$-least element of S. Then, $y \in S$ and, for every $x \prec y$, $x \notin S$. Hence, for every $x \prec y$, $P(x)$ holds. From the hypothesis of the theorem, $P(y)$ holds. That is, $y \notin S$. This is a contradiction. $\qquad\square$

Definition 3.7 f is a *partial function* from A to B iff there exists $A' \subseteq A$ such that $f : A' \to B$.

Theorem 3.8 (Recursive definitions) *Let $(A, \prec)$ be a wellordering. Let*

$$F : A \times P \to B$$

be a function where B is a set and P is the set of partial functions from A to B. Then there is a unique function $G : A \to B$ such that

$$G(y) = F\left(y, G \restriction \{x \mid x \prec y\}\right)$$

for every $y \in A$.

The point is that the equation in Theorem 3.8 determines the function G. We describe this as a *recursive definition* of G because the equation is a recipe for finding $G(y)$ based on the earlier values $G(x)$ for $x \prec y$. Specifically, the function F outputs $G(y)$ when we input y and the restriction of G to $\{x \mid x \prec y\}$.

The set P in the Theorem 3.8 might seem a bit mysterious at first. Not all partial functions are relevant to the statement and proof of the theorem. Really, we only use that, for every $y \in A$,

$$G \restriction \{x \mid x \prec y\} \in P.$$

But there would be no advantage in shrinking down P to just the partial functions we need.

Proof of Theorem 3.8 Let us say that g is a *z-approximation* iff

- $z \in A$,
- g is a partial function from A to B,

- $\text{dom}(g) = \{y \mid y \preccurlyeq z\}$ and
- for every $y \preccurlyeq z$,

$$g(y) = F\left(y, g \upharpoonright \{x \mid x \prec y\}\right).$$

The reason we call it a z-approximation is that g approximates G up to and including z.

Claim 3.8.1 *For every $z \in A$, there exists at most one z-approximation.*

Proof of claim Fix $z \in A$. Let g and h be z-approximations. We prove that $g(y) = h(y)$ by induction on $y \preccurlyeq z$ using Theorem 3.6. Let $y \preccurlyeq z$. Our induction hypothesis is that, for every $x \prec y$,

$$g(x) = h(x).$$

In other words,

$$g \upharpoonright \{x \mid x \prec y\} = h \upharpoonright \{x \mid x \prec y\}.$$

Therefore,

$$g(y) = F\left(y, g \upharpoonright \{x \mid x \prec y\}\right) = F\left(y, h \upharpoonright \{x \mid x \prec y\}\right) = h(y),$$

which completes the proof of the claim. $\square$

Claim 3.8.2 *For every $z \in A$, there exists a z-approximation.*

Proof of claim We argue by induction using Theorem 3.6. Let $z \in A$. Our induction hypothesis is that, for every $y \prec z$, there exists a y-approximation; call it g_y. By Claim 3.8.1, g_y is the only y-approximation. Observe that if $x \prec y \prec z$, then

$$g_y \upharpoonright \{w \mid w \preccurlyeq x\}$$

is an x-approximation. Since there is only one x-approximation,

$$g_x = g_y \upharpoonright \{w \mid w \preccurlyeq x\}$$

whenever $x \prec y \prec z$. Let

$$h = \bigcup_{y \prec z} g_y.$$

Then h is a partial function from A to B with

$$\text{dom}(h) = \bigcup_{y \prec z} \{x \mid x \preccurlyeq y\} = \{x \mid x \prec z\}$$

and, for every $y \prec z$,

$$g_y = h \restriction \{x \mid x \preccurlyeq y\}.$$

Now let

$$g_z = h \cup \{(z, F(z, h))\}.$$

In other words,

$$g_z(y) = \begin{cases} h(y) & \text{if } y \prec z \\ F(z, h) & \text{if } y = z. \end{cases}$$

It is easy to see that g_z is a z-approximation.   □

By the two claims, for each $z \in A$, there is a unique z-approximation, which we call g_z. As we already calculated, if $y \prec z$, then

$$g_y = g_z \restriction \{x \mid x \preccurlyeq y\}.$$

Define

$$G = \bigcup_{z \in A} g_z.$$

Then G is a function with

$$\mathrm{dom}(G) = \bigcup_{z \in A} \{y \mid y \preccurlyeq z\} = A$$

and, for every $z \in A$,

$$G(z) = g_z(z) = F\left(z, g_z \restriction \{y \mid y \prec z\}\right) = F\left(z, G \restriction \{y \mid y \prec z\}\right).$$

Thus G witnesses the conclusion of Theorem 3.8. To see that G is the unique witness to the theorem, argue just like in the proof of Claim 3.8.1.   □

3.2 Ordinal numbers

We start by repeating a definition from Exercise 2.3.

Definition 3.9 A is a *transitive* set iff for all x and y, if $x \in y$ and $y \in A$, then $x \in A$.

Equivalently, A is a transitive set iff for every $y \in A$, $y \subseteq A$. The importance of transitivity will not really begin to be apparent until Definition 3.15. First, here are three basic facts about how certain operations preserve transitivity.

Lemma 3.10 *Let A be a transitive set. Then $A \cup \{A\}$ is also a transitive set.*

Proof Let $y \in A \cup \{A\}$ and $x \in y$. We must show that $x \in A \cup \{A\}$. In fact, we will show that $x \in A$. Either $y \in A$ or $y \in \{A\}$. If $y \in A$, then $x \in A$ since A is transitive. If $y \in \{A\}$, then $y = A$, so $x \in A$. □

Lemma 3.11 *If A is a transitive set, then so is $\mathcal{P}(A)$.*

Proof Assume that A is transitive, $z \in \mathcal{P}(A)$ and $y \in z$. We must show that $y \in \mathcal{P}(A)$. By our assumption, $z \subseteq A$ and $y \in z$, so $y \in A$. Since A is transitive, $y \subseteq A$. Thus, $y \in \mathcal{P}(A)$ as desired. □

Lemma 3.12 *If $\mathcal{F}$ is a family of transitive sets, then $\bigcup \mathcal{F}$ is a transitive set too.*

Proof Let $y \in \bigcup \mathcal{F}$. Pick $A \in \mathcal{F}$ such that $y \in A$. Since A is transitive, $y \subseteq A$. Hence $y \subseteq \bigcup \mathcal{F}$ as desired. □

As motivation for the next definition, observe that membership is not a relation. This is because $\{(x, y) \mid x \in y\}$ is not a set. For if it were a set, then it would be a relation and its domain would also be a set. But its domain would be $\{x \mid x = x\}$, which is not a set by Exercise 2.6. On the other hand, if A is a set, then so is

$$\{(x, y) \in A \times A \mid x \in y\}$$

by the fact that $A \times A$ is a set and the Comprehension Scheme.

Definition 3.13 When we write $(A, \in)$ we really mean

$$(A, \{(x, y) \in A \times A \mid x \in y\}).$$

Similarly, when we say that $(A, \in)$ *is transitive*, we really mean that the restriction of $\in$ to A is a transitive relation on A.

Do not get confused between A being transitive and $(A, \in)$ being transitive. The meaning is different. For example, if $A = \{\{x\}\}$, then $\in$ is a transitive relation when restricted to A because A only has one element, but A is not a transitive set because $\{x\} \in A$ and $\{x\} \not\subseteq A$. The two notions of transitivity are related though.

Lemma 3.14 *Suppose that A is a transitive set. Then $(A, \in)$ is transitive iff for every $z \in A$, z is a transitive set.*

Proof Given that A is a transitive set, the following six statements are equivalent.

1. $(A, \in)$ is transitive.
2. $\forall\, x, y, z \in A\ ((x \in y \text{ and } y \in z) \implies x \in z)$.
3. $\forall\, z \in A\ \forall\, y \in z \cap A\ \forall\, x \in y \cap A\ (x \in z)$.
4. $\forall\, z \in A\ \forall\, y \in z\ \forall\, x \in y \cap A\ (x \in z)$.
5. $\forall\, z \in A\ \forall\, y \in z\ \forall\, x \in y\ (x \in z)$.
6. $\forall\, z \in A\ (z \text{ is a transitive set})$.

The first equivalence $(1 \iff 2)$ is by definition. The second $(2 \iff 3)$ is just logic. The third $(3 \iff 4)$ is because if $z \in A$, then $z \subseteq A$ since A is a transitive set, so $z \cap A = z$. The fourth $(4 \iff 5)$ is similar: if $z \in A$ and $y \in z$, then $y \in A$ (since A is a transitive set), hence $y \subseteq A$ (again since A is a transitive set), so $y \cap A = y$. The fifth $(5 \iff 6)$ is by definition. $\qquad\square$

Now we come to one of the most important definitions of the course.

Definition 3.15 A set α is an *ordinal* iff α is a transitive set and $(\alpha, \in)$ is a wellordering.

By the Foundation Axiom, α is an ordinal iff α is a transitive set and $(\alpha, \in)$ is a strict linear ordering. Also, by the Foundation Axiom, $(\alpha, \in)$ is always irreflexive. Combining these observations with Lemma 3.14 we get the following useful characterization of when a set is an ordinal.

Lemma 3.16 *A set α is an ordinal iff α is a transitive set, every element of α is a transitive set and $(\alpha, \in)$ is total.*

In fact, we have seen some ordinals. If $n \in \omega$, then n is an ordinal. Also, ω is an ordinal. We have also seen transitive sets that are not ordinals. For example, for every $n \in \omega$, if $n > 2$, then V_n is a transitive set that is not an ordinal.

It is important to have a reasonably good picture of where we are headed before plunging into technical facts about ordinals. As you read this paragraph, beware of the significant work required to justify this description of the ordinals, work that is captured by Lemmas 3.18, 3.19, 3.20, 3.21 and 3.22, and results on ordinal addition in the next section. Here is the picture you should have in mind. Starting from the empty set, we use the operation

$$\alpha \mapsto \alpha \cup \{\alpha\}$$

at successor stages and take unions at limit stages to generate all the ordinals beginning with the natural numbers $0 = \emptyset$, $1 = \{0\}$, $2 = \{0, 1\}$, $3 = \{0, 1, 2\}$, etc. The next ordinal after all the natural numbers is the set of natural numbers,

$$\omega = \{0, 1, 2, \ldots\}.$$

After ω comes the infinite sequence of ordinals

$$\omega + 1 = \{0, 1, 2, \ldots, \omega\}$$
$$\omega + 2 = \{0, 1, 2, \ldots, \omega, \omega + 1\}$$
$$\omega + 3 = \{0, 1, 2, \ldots, \omega, \omega + 1, \omega + 2\}$$

$$\vdots$$

followed by infinitely more ordinals

$$\omega + \omega = \{0, 1, 2, \ldots, \omega, \omega + 1, \omega + 2, \ldots\}$$
$$\omega + \omega + 1 = \{0, 1, 2, \ldots, \omega, \omega + 1, \omega + 2, \ldots, \omega + \omega\}$$
$$\omega + \omega + 2 = \{0, 1, 2, \ldots, \omega, \omega + 1, \omega + 2, \ldots, \omega + \omega, \omega + \omega + 1\}$$

$$\vdots$$

after which comes the ordinal

$$\omega + \omega + \omega.$$

Skipping ahead, we eventually get to

$$\omega + \omega + \omega + \omega$$

and, somewhat later, to

$$\omega \cdot \omega = \omega + \cdots + \omega + \cdots.$$

The list of ordinals never ends. Notice that, for natural numbers, the membership relation $\in$ coincides with the usual strict linear ordering $<$ on the natural numbers, and that this pattern continues through the ordinals we have listed above. Namely,

$$0 \in 1 \in 2 \in 3 \in \cdots \in \omega \in \omega + 1 \in \omega + 2 \in \omega + 3 \in \cdots.$$

Returning now to a rigorous exposition, we make the following general notational rules.

Definition 3.17 If α and β are ordinals, then we may write

$$\alpha < \beta \iff \alpha \in \beta$$

and

$$\alpha \leq \beta \iff (\alpha < \beta \text{ or } \alpha = \beta).$$

This convention will allow us to write

$$0 < 1 < 2 < 3 < \cdots < \omega < \omega + 1 < \omega + 2 < \omega + 3 < \cdots$$

with the same meaning as

$$0 \in 1 \in 2 \in 3 \in \cdots \in \omega \in \omega + 1 \in \omega + 2 \in \omega + 3 \in \cdots$$

once we finish developing the theory of ordinals.

Next are some important basic principles about ordinals to which we have alluded above. The proofs might seem confusing the first time through because they involve all three notions: $\in$, $<$ and $\subseteq$. Read slowly and attentively. Remember, in general, once we establish that α and β are ordinals (but not before) we are free to write $\alpha < \beta$ instead of $\alpha \in \beta$.

Lemma 3.18 *Let β be an ordinal. Then every element of β is an ordinal. Thus*

$$\beta = \{\alpha \mid \alpha \text{ is an ordinal and } \alpha < \beta\}.$$

Proof Let $\alpha \in \beta$. By the forward direction of Lemma 3.14, every member of β is a transitive set. Thus α is a transitive set. Because β is a transitive set, $\alpha \subseteq \beta$. Since $(\beta, \in)$ is a wellordering, so is $(\alpha, \in)$. (Every subset of a wellordered set is also wellordered.) Therefore, α is an ordinal by Definition 3.15. $\square$

Lemma 3.19 *Let γ and δ be ordinals. Then*

$$\gamma \leq \delta \iff \gamma \subseteq \delta.$$

Proof First we prove the forward (left to right) direction. Assume $\gamma \leq \delta$. If $\gamma = \delta$, we are done. So assume $\gamma < \delta$. In other words, $\gamma \in \delta$. Since δ is a transitive set, $\gamma \subseteq \delta$. So, again, we are done.

Now, for the proof of the reverse (right to left) direction, assume that $\gamma \subseteq \delta$. If $\gamma = \delta$, then we are done, so assume that $\gamma \neq \delta$.

Then $\delta - \gamma$ (set difference) is a non-empty subset of δ and $(\delta, <)$ is a wellordering, so we may let β be the $<$-least element of $\delta - \gamma$.

Claim 3.19.1 $\beta = \gamma$.

Proof of claim By the Extensionality Axiom, Lemma 3.18 and the fact that both β and γ are subsets of δ, the claim is equivalent to saying that, for every $\alpha < \delta$,

$$\alpha < \beta \iff \alpha < \gamma.$$

Let $\alpha < \delta$ be given. If $\alpha < \beta$, then $\alpha < \gamma$ by the definition of β. In order to finish the proof of the claim, we assume that $\alpha < \gamma$ but $\alpha \not< \beta$ and work towards a contradiction. Since α and β are both elements of δ and $(\delta, <)$ is a total ordering, saying that $\alpha \not< \beta$ is equivalent to saying that $\alpha \geq \beta$. Combining facts, we have that

$$\beta \leq \alpha < \gamma.$$

This is the same as saying that either

$$\beta \in \alpha \text{ and } \alpha \in \gamma$$

or else

$$\beta = \alpha \text{ and } \alpha \in \gamma.$$

Either way, since γ is a transitive set, we conclude that $\beta \in \gamma$. But $\beta \notin \gamma$ by the definition of β. This contradiction proves the claim. $\square$

Now we are done proving the reverse direction of Lemma 3.19 because $\gamma = \beta < \delta$. $\square$

Lemma 3.20 *If β and γ are ordinals, then either $\beta < \gamma$ or $\beta = \gamma$ or $\beta > \gamma$.*

Proof Assume that $\gamma \not< \beta$. By Lemma 3.19, $\gamma \not\subseteq \beta$. Because $(\gamma, <)$ is a wellordering, we may let α be the $<$-least element of $\gamma - \beta$ (set difference). Then $\alpha \subseteq \beta$ but $\alpha \notin \beta$. Hence, $\alpha \leq \beta$ by Lemma 3.19 but $\alpha \not< \beta$. Thus $\beta = \alpha < \gamma$. $\square$

Lemma 3.21 *Let α be an ordinal and $\beta = \alpha \cup \{\alpha\}$. Then β is an ordinal and $\alpha < \beta$. Moreover, if γ is an ordinal and $\alpha < \gamma$, then $\beta \leq \gamma$.*

Proof By Lemma 3.10, β is a transitive set. Obviously, every element of β is also a transitive set and $(\beta, \in)$ is total. By Lemma 3.16, β is an ordinal. Clearly, $\alpha < \beta$. Suppose that γ is an ordinal and $\alpha < \gamma$. Then $\alpha \subseteq \gamma$ since γ is transitive. Hence $\beta = \alpha \cup \{\alpha\} \subseteq \gamma$. By Lemma 3.19, $\beta \leq \gamma$.   □

Because of Lemma 3.21, it is natural to write $\alpha + 1 = \alpha \cup \{\alpha\}$ for ordinals α. We call $\alpha + 1$ a *successor* ordinal. Non-zero ordinals that are not successor ordinals are called *limit ordinals*.

Lemma 3.22   *Let A be a set of ordinals and $\beta = \bigcup A$. Then β is an ordinal and, for every $\alpha \in A$, $\alpha \leq \beta$. Moreover, if γ is an ordinal and $\alpha \leq \gamma$ for every $\alpha \in A$, then $\beta \leq \gamma$.*

The proof of Lemma 3.22 is left as a practice problem; it builds on the chain of lemmas starting from Lemma 3.12. Recall that *supremum* is another way to say *least upper bound*. Lemma 3.22 says that if A is a set of ordinals, then $\sup(A) = \bigcup A$. If A is a set of ordinals and $\sup(A) \in A$, then $\sup(A)$ is the *maximum* element of A, in which case we may write $\max(A) = \sup(A) = \bigcup A$. But not every set of ordinals has a maximum element. For example,

$$\sup\left(\{5, 6, 7, \dots\}\right) = \bigcup \{5, 6, 7, \dots\} = \omega$$

but

$$\omega \notin \{5, 6, 7, \dots\},$$

so $\{5, 6, 7, \dots\}$ does not have a maximum element. On the other hand,

$$\sup\left(\{5, 6, 7, \dots, \omega\}\right) = \bigcup \{5, 6, 7, \dots, \omega\} = \omega$$

and

$$\omega \in \{5, 6, 7, \dots, \omega\},$$

so

$$\max\left(\{5, 6, 7, \dots, \omega\}\right) = \omega.$$

If A is a set of ordinals and $A \neq \emptyset$, then A has a $<$-least element called *the minimum of A* and denoted $\min(A)$. To justify the definition of $\min(A)$ use the fact that $A \subseteq \sup(A) + 1$ and $(\sup(A) + 1, <)$ is a wellordering.

Since they are important, let us state versions of Theorems 3.6 and 3.8 that relate to ordinals.

Theorem 3.23 (Proofs by induction)   *Let $P(\alpha)$ be a statement about a variable α. Assume that, for every ordinal β,*

$$(\forall \alpha < \beta \; P(\alpha) \text{ holds}) \implies P(\beta) \text{ holds}.$$

Then, for every ordinal γ, $P(\gamma)$ holds.

Proof   Consider an arbitrary ordinal γ and let $\theta = \gamma + 1$. We prove that $P(\beta)$ holds for every $\beta < \theta$ by induction along the wellordering $(\theta, <)$ using Theorem 3.6. But there is really nothing to do because the assumption here is at least as strong as the assumption of Theorem 3.6. To see the connection more clearly, substitute $(A, \prec) = (\theta, <)$, $y = \beta$ and $x = \alpha$ in the statement of Theorem 3.6.   □

Notice that Theorem 3.23 gives us a method for showing that a property holds for every ordinal, not just every ordinal up to a given ordinal. Often, the verification of the hypothesis of Theorem 3.23 is broken up into three cases: $\beta = 0$, $\beta = \alpha + 1$ and β a limit ordinal. Similarly, in applications of Theorem 3.24 below, sometimes the definition of $F(\beta, \cdot)$ is broken up into three cases: $\beta = 0$, $\beta = \alpha + 1$ and β a limit ordinal.

Theorem 3.24 (Recursive definitions)   *Suppose that θ is an ordinal. Let*

$$F : \theta \times P \to B$$

be a function where B is a set and P is the set of partial functions from θ to B. Then there is a unique function $G : \theta \to B$ such that

$$G(\beta) = F(\beta, G \upharpoonright \beta)$$

for every $\beta < \theta$.

Notice that Theorem 3.24 has a different flavor than Theorem 3.23 in that it only tells us how to make recursive definitions up to a given ordinal θ. Although it is possible to state a theorem on recursive definitions through all the ordinals, the statement would be overly technical, so we prefer to show how such definitions can be made and justified by giving an example.

Here is an illustration of a recursive definition followed by a proof by induction on the ordinals. If θ is an ordinal, then using Theorem 3.24, we may define a function $\alpha \mapsto V_\alpha$ with domain θ

by saying that

$$V_0 = \emptyset,$$

$$V_{\alpha+1} = \mathcal{P}\left(V_\alpha\right)$$

and

$$V_\beta = \bigcup \{V_\alpha \mid \alpha < \beta\}$$

if β is a limit ordinal. If we really want to match up this definition of V_α with Theorem 3.24, then we end up with $G : \alpha \mapsto V_\alpha$ starting with

$$F(\beta, g) = \begin{cases} \emptyset & \text{if } \beta = 0 \\ \mathcal{P}(g(\alpha)) & \text{if } \beta = \alpha + 1 \\ \bigcup \{g(\alpha) \mid \alpha < \beta\} & \text{if } \beta \text{ is a limit ordinal.} \end{cases}$$

Notice that the definition V_α does not depend on θ so really what we have is a way of assigning a set V_α to each ordinal α. The assignment $\alpha \mapsto V_\alpha$ is not a function because what should be its domain is not a set by the following lemma.

Lemma 3.25   *There is no set of all ordinals.*

Proof   Suppose for contradiction that

$$\Omega = \{\alpha \mid \alpha \text{ is an ordinal}\}$$

is a set. It is easy to see that Ω is transitive and $(\Omega, \in)$ is a wellordering. But then Ω is an ordinal. In other words, $\Omega \in \Omega$. But then

$$\cdots \in \Omega \in \Omega \in \Omega$$

is an infinite descending sequence of members of Ω. This contradicts that $(\Omega, \in)$ is a wellordering.   □

Notice that, once we realized that $\Omega \in \Omega$, we could have used the Foundation Axiom to finish the proof of Lemma 3.25 slightly more quickly. The Foundation Axiom automatically holds for ordinals, which is essentially how we got away without using it above.

We have been discussing recursion; now let us see an example of induction.

Lemma 3.26   *Let δ be an ordinal and $\beta < \delta$. Then the following hold.*

(1)$_\delta$ V_δ *is a transitive set.*
(2)$_\delta$ *For every* $\beta < \delta$, $V_\beta \subseteq V_\delta$.

Proof We prove the lemma by induction on δ. The induction hypothesis is that the lemma holds for every $\gamma < \delta$.

Base case $\delta = 0$.

In this case, (1)$_0$ holds because $V_0 = \emptyset$ is a transitive set, whereas (2)$_0$ holds because there is nothing to check.

Successor case $\delta = \gamma + 1$.

In this case, $V_\delta = \mathcal{P}(V_\gamma)$. By Lemma 3.11, since (1)$_\gamma$ holds, (1)$_\delta$ holds too. Also, since (1)$_\gamma$ holds, if $y \in V_\gamma$, then $y \subseteq V_\gamma$, so $y \in V_\delta$. This shows that $V_\gamma \subseteq V_\delta$. This conclusion together with (2)$_\gamma$ implies that (2)$_\delta$ holds.

Limit case δ *is a limit ordinal.*

In this case, $V_\delta = \bigcup\{V_\gamma \mid \gamma < \delta\}$. The fact that (1)$_\delta$ holds is immediate from Lemma 3.12 and the assumption that (1)$_\gamma$ for $\gamma < \delta$. Also, (2)$_\delta$ follows from (2)$_\gamma$ for $\gamma < \delta$. $\square$

Here is an interesting fact that plays no role in the rest of the book. It turns out that the Foundation Axiom is equivalent to the statement

$$\forall x \, \exists \alpha \, (\alpha \text{ is an ordinal and } x \in V_\alpha).$$

That is, every set belongs to some V_α. The proof can be found in various graduate level textbooks on set theory.

Now we continue developing the theory of ordinal numbers.

Definition 3.27 Let $R \subseteq A \times A$ and $S \subseteq B \times B$. Suppose that $\pi : A \to B$. Then π is an *isomorphism from* (A, R) *to* (B, S) iff π is a bijection from A to B and, for every $x, y \in A$,

$$xRy \iff \pi(x)S\pi(y).$$

We write $\pi : (A, R) \simeq (B, S)$ in this case.

The relationship between ordinals and arbitrary wellorderings is summarized by the following theorem, which is a special case of a more general theorem by Mostowski.

Theorem 3.28 (Mostowski collapse)   *Let $(A, \prec)$ be a wellordering. Then there exists an ordinal α and an isomorphism*

$$\pi : (A, \prec) \simeq (\alpha, <).$$

Moreover, α and π are unique in the sense that if α' is an ordinal and

$$\pi' : (A, \prec) \simeq (\alpha', <)$$

is an isomorphism, then $\alpha' = \alpha$ and $\pi' = \pi$.

The isomorphism π in the conclusion of Theorem 3.28 is called the *Mostowski collapse of* $(A, \prec)$. Here is a specific example. Let

$$A = \omega - \{0, 1, 4\}$$

and

$$m \prec n \iff (m, n \in A \text{ and } m < n).$$

Then the Mostowski collapse $\pi : (A, \prec) \simeq (\omega, <)$ is determined by the following table.

m	$\pi(m)$
2	0
3	1
5	2
6	3
7	4
$\vdots$	$\vdots$

In general, if x is the $\prec$-least element of A, then $\pi(x) = 0$. If, in addition, y is the $\prec$-least element of $A - \{x\}$, then $\pi(y) = 1$. And so on. Notice that the list on the left has gaps (0, 1 and 4 are missing) but the list on the right has no gaps. Intuitively, the function π is called the *collapse* because it gets rid of the gaps.

Another way to think about the Mostowski collapse is as follows. Suppose we list A as $a_0, a_1, \ldots$ in increasing order according to $\prec$ using ordinals as indices. In the example above, we would have $a_0 = 2$, $a_1 = 3$, $a_2 = 5$, $a_3 = 6$, etc. What we would end up with is, for every $b \in A$,

$$b = a_{\pi(b)}.$$

In other words, the Mostowski collapse tells us the index.

Proof of Theorem 3.28 We are given a wellordering $(A, \prec)$ and we are looking for an ordinal α and an isomorphism

$$\pi : (A, \prec) \simeq (\alpha, <).$$

Let us work backwards to see what the definitions of α and π must be. Suppose we already have α and π as above. Then, for all $x, y \in A$,

$$x \prec y \iff \pi(x) < \pi(y) \iff \pi(x) \in \pi(y).$$

Therefore, for every $y \in A$,

$$\pi(y) = \{\pi(x) \mid x \prec y\}.$$

Notice that this is a recursive equation because the value of $\pi(y)$ is determined from $\pi \restriction \{x \mid x \prec y\}$. By Theorem 3.8, there is at most one function π that satisfies this recursive equation. Moreover, once we know π, we also know α because $\alpha = \pi[A]$. This explains why α and π are unique. In other words, it proves the *moreover* part of Theorem 3.28. It also gives us the clue we need to prove the *existence* part of Theorem 3.28, which is what we do next.

Without assuming that α is an ordinal and π is an isomorphism, recursively define $\pi : A \to B$ by setting

$$\pi(y) = \{\pi(x) \mid x \prec y\}.$$

This is not entirely legitimate, at least not if we wish to implement Theorem 3.8, because we are required to name the set B before we define π by recursion. We will leave out the argument that we can choose B in advance since it involves reasoning from the axioms of ZFC that is overly technical from the perspective of this book.

Claim 3.28.1 π *is an isomorphism from* $(A, \prec)$ *to* $(\pi[A], \in)$.

Proof First we show that π is an injection. Consider arbitrary $x, y \in A$ such that $x \neq y$. Then either $x \prec y$ or $y \prec x$. Suppose that $x \prec y$. Then $\pi(x) \in \pi(y)$ by the definition of π. On the other hand, $\pi(x) \notin \pi(x)$ by the Foundation Axiom. Hence $\pi(x) \neq \pi(y)$. The proof is similar if $y \prec x$.

Every injection is a bijection with its range. It remains to see that π is order preserving. We already noted that if $x \prec y$, then $\pi(x) \in \pi(y)$. For the converse, suppose that $\pi(x) \in \pi(y)$. Then there exists $x' \prec y$ such that $\pi(x') = \pi(x)$. Since π is an injection, $x' = x$. Hence $x \prec y$ as required. $\square$

Claim 3.28.2 $\pi[A]$ *is an ordinal.*

Proof If $u \in v$ and $v \in \pi[A]$, then there are $x, y \in A$ such that $u = \pi(x)$ and $v = \pi(y)$, so $u \in \pi[A]$. This shows that $\pi[A]$ is a transitive set. By the previous claim, $(\pi[A], \in)$ is an isomorphic copy of $(A, \prec)$. Since $(A, \prec)$ is a wellordering, so is $(\pi[A], \in)$. $\square$

Setting $\alpha = \pi[A]$ concludes the proof of Theorem 3.28. $\square$

The previous theorem justifies the following definition.

Definition 3.29 Let $(A, \prec_A)$ be a wellordering. Then

$$\text{type}(A, \prec_A)$$

is the unique ordinal isomorphic to $(A, \prec_A)$.

Suppose that A is a set of ordinals. Say $A \subseteq \beta$. Then $(\beta, <)$ is a wellordering and hence so is $(A, <)$. This motivates the following notation.

Definition 3.30 Let A be a set of ordinals. Then

$$\text{type}(A) = \text{type}(A, <).$$

We end this section with a technical lemma about the Mostowski collapse of a set of ordinals. It will be used in Chapter 4.

Lemma 3.31 *Let A be a set of ordinals and suppose that $A \subseteq \beta$. Let*

$$\alpha = \text{type}(A)$$

and

$$\pi : (A, <) \simeq (\alpha, <)$$

be the Mostowski collapse of $(A, <)$. Then, for every $\eta \in A$,

$$\pi(\eta) \le \eta.$$

Moreover, $\alpha \le \beta$.

Proof First we prove that $\pi(\eta) \le \eta$ by induction on $\eta \in A$. Recall that if ζ and η are ordinals, then

$$\zeta \le \eta \iff \zeta \subseteq \eta.$$

This is by Lemma 3.19. Therefore, what we must show is equivalent to

$$\pi(\eta) \subseteq \eta$$

for every $\eta \in A$. By the definition of the Mostowski collapse,

$$\pi(\eta) = \{\pi(\zeta) \mid \zeta \in A \cap \eta\}.$$

By the induction hypothesis, for every $\zeta \in A \cap \eta$,

$$\pi(\zeta) \leq \zeta$$

so

$$\pi(\zeta) < \zeta + 1 \leq \eta.$$

Therefore,

$$\pi(\eta) \subseteq \bigcup \{\zeta + 1 \mid \zeta \in A \cap \eta\} \subseteq \eta.$$

This completes the proof by induction. In particular, we have seen that if $\eta \in A$, then

$$\pi(\eta) \leq \eta < \beta.$$

Since $\alpha = \pi[A]$, this implies that $\alpha \subseteq \beta$, hence $\alpha \leq \beta$. $\qquad\square$

3.3 Ordinal arithmetic

In this section, we extend the usual notions of addition, multiplication and exponentiation of natural numbers to all ordinals.

There are various ways to join a pair of sets. For example, given A and B, we can form their union $A \cup B$. But sometimes we want to take a *disjoint union* instead. This means that we take the union of disjoint copies of A and B. The advantage is that, given a point in the disjoint union, it comes either from A or from B but not both. A convenient way to define the disjoint union is $(\{0\} \times A) \cup (\{1\} \times B)$. Building on this idea, sometimes we are given two wellorderings and we want to put one after the other to form a new wellordering. The following definition tells us how.

Definition 3.32 Let $(A, \prec_A)$ and $(B, \prec_B)$ be wellorderings. Then their *concatenation* is

$$(A, \prec_A)^\frown (B, \prec_B) = (C, \prec_C)$$

where

$$C = (\{0\} \times A) \cup (\{1\} \times B)$$

and, for all $(i, x), (j, y) \in C$,

$$(i, x) \prec_C (j, y) \iff \begin{pmatrix} (i = j = 0 \text{ and } x \prec_A y) \\ \text{or} \\ (i = 0 \text{ and } j = 1) \\ \text{or} \\ (i = j = 1 \text{ and } x \prec_B y) \end{pmatrix}.$$

Lemma 3.33 *Let $(A, \prec_A)$ and $(B, \prec_B)$ be wellorderings, and*

$$(C, \prec_C) = (A, \prec_A) ^\frown (B, \prec_B).$$

Then $(C, \prec_C)$ is a wellordering.

Proof It is straightforward to verify that $(C, \prec_C)$ is a strict linear ordering. For contradiction, suppose that $\langle (i_n, x_n) \mid n < \omega \rangle$ is an infinite descending sequence from C. Since $(B, \prec_B)$ is a wellordering, there is some $m < \omega$ such that, for every $n < \omega$, if $m < n$, then $i_n = 0$. But then $\langle x_n \mid n < \omega - m \rangle$ is an infinite descending sequence from A, which is a contradiction. □

Definition 3.34 Let α and β be ordinals. Then their *sum* is

$$\alpha + \beta = \text{type}\left((\alpha, <) ^\frown (\beta, <) \right).$$

In other words, $\alpha + \beta$ is the unique ordinal isomorphic to

$$(\alpha, <) ^\frown (\beta, <).$$

It is the combination of Lemma 3.33 and Theorem 3.28 that justifies Definition 3.34. We also remark that, earlier, we defined $\alpha + 1$ to mean $\alpha \cup \{\alpha\}$ whereas here we used a different definition of $\alpha + 1$. The two definitions coincide because

$$(\alpha, <) ^\frown (1, <) \simeq (\alpha \cup \{\alpha\}, <).$$

Make sure you understand what is being asserted here and why it is true!

Example $3 + 2 = 5$. This is because if $(C, \prec_C) = (3, <) ^\frown (2, <)$, then

$$(0, 0) \prec_C (0, 1) \prec_C (0, 2) \prec_C (1, 0) \prec_C (1, 1)$$

and so we see that $(C, \prec_C) \simeq (5, <)$.

Example $2 + 3 = 5$. This is because if $(C, \prec_C) = (2, <)^\frown(3, <)$, then

$$(0, 0) \prec_C (0, 1) \prec_C (1, 0) \prec_C (1, 1) \prec_C (1, 2)$$

and so we see that $(C, \prec_C) \simeq (5, <)$.

Example $3 + \omega = \omega$. This is because $(C, \prec_C) = (3, <)^\frown(\omega, <)$ consists of the initial segment

$$(0, 0) \prec_C (0, 1) \prec_C (0, 2)$$

followed by the infinite tail

$$(1, 0) \prec_C (1, 1) \prec_C (1, 2) \prec_C (1, 3) \prec_C (1, 4) \prec_C (1, 5) \prec_C \cdots$$

from which we see that

$$(C, \prec_C) \simeq (\omega, <)$$

according to the isomorphism

$$(i, n) \mapsto \begin{cases} n & \text{if } i = 0 \\ 3 + n & \text{if } i = 1. \end{cases}$$

Notice that $3 + \omega = \omega \neq \omega + 3$, so ordinal addition is not commutative! However, ordinal addition is associative.

Lemma 3.35 *For all ordinals α, β and γ,*

$$(\alpha + \beta) + \gamma = \alpha + (\beta + \gamma).$$

Proof Let

$$(C, \prec_C) = (\alpha, <)^\frown(\beta, <)$$

and

$$D = (C, \prec_C)^\frown(\gamma, <).$$

Let

$$(E, \prec_E) = (\beta, <)^\frown(\gamma, <)$$

and

$$F = (\alpha, <)^\frown(E, \prec_E).$$

It is enough to see that there is an isomorphism

$$\pi : (D, \prec_D) \simeq (F, \prec_F).$$

Define π by cases according to the following list of equations.

$$\pi((0, (0, \xi))) = (0, \xi)$$
$$\pi((0, (1, \xi))) = (1, (0, \xi))$$
$$\pi((1, \xi)) = (1, (1, \xi))$$

It is clear that this works.   □

Addition can also be defined recursively in terms of the assignment that takes an ordinal α to its successor $\alpha + 1$. The following lemma shows this.

Lemma 3.36   *Let α and θ be ordinals. Then there is a unique function f with domain θ such that, for every $\gamma < \theta$,*

$$f(\gamma) = \begin{cases} \alpha & \textit{if } \gamma = 0 \\ f(\beta) + 1 & \textit{if } \gamma = \beta + 1 \\ \sup\left(\{f(\beta) \mid \beta < \gamma\}\right) & \textit{if } \gamma \textit{ is a limit ordinal,} \end{cases}$$

namely, the function given by

$$f(\gamma) = \alpha + \gamma.$$

Sketch of proof   Use induction on $\gamma < \theta$ to see that that ordinal addition satisfies the three conditions we specified for f. Namely,

- $\alpha + 0 = \alpha$,
- $\alpha + (\beta + 1) = (\alpha + \beta) + 1$ and
- if γ is a limit ordinal, then $\alpha + \gamma = \sup\left(\{\alpha + \beta \mid \beta < \gamma\}\right)$.

Then apply Theorem 3.24. The details comprise Exercise 3.5.   □

Here is an entertaining false argument. We saw that

$$3 + \omega = \omega = 0 + \omega.$$

Subtracting ω from both sides of this equation, we see that $3 = 0$. Do you see why this is nonsense? There is no inverse operation for ordinal addition. The following lemma is as close as we come to ordinal subtraction.

Lemma 3.37   *Let $\alpha \leq \beta$ be ordinals. Then there is a unique ordinal δ such that $\alpha + \delta = \beta$.*

Sketch of proof   Let $D = \{\eta \mid \alpha \le \eta < \beta\}$. It is clear that

$$(\alpha, <)^\frown (D, <) \simeq (\beta, <).$$

By Theorem 3.28, there is a unique ordinal δ such that

$$(D, <) \simeq (\delta, <),$$

namely,

$$\delta = \text{type}(D).$$

It is clear that

$$(\alpha, <)^\frown (\delta, <) \simeq (\beta, <).$$

This is equivalent to saying that

$$\alpha + \delta = \beta.$$

It remains to see that δ is the unique solution to this equation. Suppose that

$$\alpha + \delta' = \beta.$$

Then

$$(\alpha, <)^\frown (\delta', <) \simeq (\beta, <),$$

from which one can argue that

$$(D, <) \simeq (\delta', <).$$

By the uniqueness clause of Theorem 3.28, we see that $\delta' = \delta$. The remaining details form Exercise 3.12. $\qquad\square$

Just before Lemma 3.37, we saw that $\delta + \omega = 3 + \omega$ does not imply $\delta = 3$. In other words, we cannot cancel on the right. It is a consequence of the uniqueness clause in Lemma 3.37 that we can cancel on the left. For example, if $\omega + \delta = \omega + 3$, then $\delta = 3$.

Now we work towards defining multiplication. For this, we use another way of putting together two wellorderings.

Definition 3.38   Let $(A, \prec_A)$ and $(B, \prec_B)$ be wellorderings. Then the lexicographic ordering on $A \times B$ is the relation $\prec$ such that, for all $(x, y), (x', y') \in A \times B$,

$$(x, y) \prec (x', y') \iff \begin{pmatrix} (x \prec_A x') \\ \text{or} \\ (x = x' \text{ and } y \prec_B y') \end{pmatrix}.$$

Lemma 3.39 *Let $(A, \prec_A)$ and $(B, \prec_B)$ be wellorderings. Then $(C, \prec_C)$ is a wellordering where $C = A \times B$ and $\prec_C$ is the lexicographic ordering on C.*

The proof of Lemma 3.39 is Exercise 3.7.

Definition 3.40 Let α and β be ordinals. Let γ be the unique ordinal isomorphic to the lexicographic ordering on $\alpha \times \beta$. Then the product is

$$\beta \cdot \alpha = \gamma.$$

This is not a misprint; by tradition, ordinal multiplication is read from right to left. By $\beta \cdot \alpha$, we mean α many copies of β, not the other way around, and sometimes it matters for infinite ordinals.

Example $3 \cdot 2 = 6$. This is because, if $C = 2 \times 3$ and $\prec_C$ is the lexicographic order on C, then

$$(0,0) \prec_C (0,1) \prec_C (0,2) \prec_C (1,0) \prec_C (1,1) \prec (1,2)$$

and so we see that $(C, \prec_C) \simeq (6, \in)$.

Example $2 \cdot 3 = 6$. This is because, if $C = 3 \times 2$ and $\prec_C$ is the lexicographic order on C, then

$$(0,0) \prec_C (0,1) \prec_C (1,0) \prec_C (1,1) \prec_C (2,0) \prec (2,1)$$

and so we see that $(C, \prec_C) \simeq (6, \in)$.

Example $3 \cdot \omega = \omega$. This is because, if $C = \omega \times 3$ and $\prec_C$ is the lexicographic order on C, then $(C, \prec_C)$ looks like

$$(0,0) \prec_C (0,1) \prec_C (0,2)$$

followed by

$$(1,0) \prec_C (1,1) \prec_C (1,2)$$

then

$$(2,0) \prec_C (2,1) \prec_C (2,2)$$

then

$$(3,0) \prec_C (3,1) \prec_C (3,2)$$

then

$$(4,0) \prec_C (4,1) \prec_C (4,2)$$

and so on. From this, we see that

$$(C, \prec_C) \simeq (\omega, <)$$

according to the isomorphism

$$(i, n) \mapsto 3 \cdot i + n.$$

Example $\omega \cdot 3 = \omega + \omega + \omega$. This is because, if $C = 3 \times \omega$ and $\prec_C$ is the lexicographic order on C, then $(C, \prec_C)$ looks like

$$(0,0) \prec_C (0,1) \prec_C (0,2) \prec_C (0,3) \prec_C (0,4) \prec_C \cdots$$

followed by

$$(1,0) \prec_C (1,1) \prec_C (1,2) \prec_C (1,3) \prec_C (1,4) \prec_C \cdots$$

then

$$(2,0) \prec_C (2,1) \prec_C (2,2) \prec_C (2,3) \prec_C (2,4) \prec_C \cdots.$$

From this we see that

$$(C, \prec_C) \simeq (\omega + \omega + \omega, <)$$

according to the isomorphism

$$(i, n) \mapsto \begin{cases} n & \text{if } i = 0 \\ \omega + n & \text{if } i = 1 \\ \omega + \omega + n & \text{if } i = 2, \end{cases}$$

which we could also write as

$$(i, n) \mapsto \omega \cdot i + n.$$

Notice that

$$3 \cdot \omega = \omega \neq \omega \cdot 3,$$

so ordinal multiplication is not commutative! However, ordinal multiplication is associative as the following lemma shows. The proof is Exercise 3.6(5).

Lemma 3.41 *For all ordinals α, β and γ,*

$$\gamma \cdot (\beta \cdot \alpha) = (\gamma \cdot \beta) \cdot \alpha.$$

Ordinal multiplication distributes over ordinal addition in the following way. The proof is the first part of Exercise 3.8.

Lemma 3.42 *For all ordinals α, β and γ,*

$$\alpha \cdot (\beta + \gamma) = (\alpha \cdot \beta) + (\alpha \cdot \gamma).$$

Ordinal multiplication can also be defined recursively in terms of ordinal addition as the following lemma shows. The proof is Exercise 3.9.

Lemma 3.43 *Let α and θ be ordinals. Then there is a unique function f with domain θ such that, for every $\gamma < \theta$,*

$$f(\gamma) = \begin{cases} 0 & \text{if } \gamma = 0 \\ f(\beta) + \alpha & \text{if } \gamma = \beta + 1 \\ \sup\left(\{f(\beta) \mid \beta < \gamma\}\right) & \text{if } \gamma \text{ is a limit ordinal,} \end{cases}$$

namely, the function given by

$$f(\gamma) = \alpha \cdot \gamma.$$

For ordinal addition and ordinal multiplication, we started with definitions in terms of wellorderings, then stated lemmas giving equivalent recursive definitions. For variety and because it is easier in this case, we give the recursive definition of ordinal exponentiation and leave the interpretation in terms of wellorderings to the reader as a rather tricky exercise. (Exercise 3.15 will get the reader started.)

Definition 3.44 For every ordinal α,

- $\alpha^0 = 1$,
- for every ordinal β,

$$\alpha^{\beta+1} = \alpha^\beta \cdot \alpha$$

  and
- for every limit ordinal γ,

$$\alpha^\gamma = \sup\left(\left\{\alpha^\beta \mid 0 < \beta < \gamma\right\}\right).$$

Example $3^\omega = \sup\left(\{3^n \mid n < \omega\}\right) = \omega < \omega^3$.

Example If $\beta \geq \omega^2$, then $\omega + \beta = \beta$. To see this, first note that

$$\omega^2 = \omega \cdot \omega = \omega \cdot (1 + \omega) = (\omega \cdot 1) + (\omega \cdot \omega) = \omega + \omega^2.$$

Now suppose that $\beta \geq \omega^2$. By Lemma 3.37, there is an ordinal δ such that $\beta = \omega^2 + \delta$. Thus,

$$\beta = \omega^2 + \delta = (\omega + \omega^2) + \delta = \omega + (\omega^2 + \delta) = \omega + \beta.$$

Exercises

Exercise 3.1 Let $(A, \prec_A)$ be a wellordering such that $A \neq \emptyset$. For each $y \in A$, define

$$\mathrm{pred}_{(A, \prec_A)}(y) = \{x \in A \mid x \prec_A y\}.$$

Suppose that $S \subsetneq A$ and, for all $x, y \in A$, if $y \in S$ and $x \prec_A y$, then $x \in S$. Prove that there exists $y \in A$ such that

$$S = \mathrm{pred}_{(A, \prec_A)}(y).$$

Exercise 3.2 Let $(A, \prec_A)$ and $(B, \prec_B)$ be wellorderings. Use Theorem 3.28 to prove that exactly one of the following conditions holds:

- $(A, \prec_A) \simeq (B, \prec_B)$.
- There exists $y \in B$,

$$(A, \prec_A) \simeq \mathrm{pred}_{(B, \prec_B)}(y) \text{ ordered by } \prec_B .$$

- There exists $y \in A$,

$$(B, \prec_B) \simeq \mathrm{pred}_{(A, \prec_A)}(y) \text{ ordered by } \prec_A .$$

Exercise 3.3 Prove that, for every ordinal β,

$$\{\alpha \in V_\beta \mid \alpha \text{ is an ordinal}\} = \beta.$$

As a hint, see the last part of Exercise 2.3.

Exercise 3.4 Prove that the following facts about ordinal addition hold for all ordinals α, β and γ.

1. $0 + \alpha = \alpha = \alpha + 0$.
2. $\beta \leq \alpha + \beta$.
3. If $\beta < \gamma$, then $\alpha + \beta < \alpha + \gamma$.
4. If $\alpha \leq \beta$, then $\alpha + \gamma \leq \beta + \gamma$.

Exercise 3.5 Complete the proof of Lemma 3.36.

Exercise 3.6 Prove that the following facts about ordinal multiplication hold for all ordinals α, β and γ.

1. $0 \cdot \alpha = 0 = \alpha \cdot 0$.
2. $\alpha \cdot 1 = 1 \cdot \alpha$.
3. If $0 < \alpha$ and $\beta < \gamma$, then $\alpha \cdot \beta < \alpha \cdot \gamma$.
4. If $\alpha \leq \beta$, then $\alpha \cdot \gamma \leq \beta \cdot \gamma$.
5. $(\alpha \cdot \beta) \cdot \gamma = \alpha \cdot (\beta \cdot \gamma)$. (That is, prove Lemma 3.41.)

Exercise 3.7 Prove Lemma 3.39.

Exercise 3.8 This exercise is on distributive laws for ordinal addition and multiplication.

1. Prove Lemma 3.42.
2. Give an example of ordinals α, β and γ such that

$$(\alpha + \beta) \cdot \gamma \neq (\alpha \cdot \gamma) + (\beta \cdot \gamma).$$

Exercise 3.9 Prove Lemma 3.43.

Exercise 3.10 Prove that the following facts about ordinal exponentiation hold for all ordinals α, β and γ.

1. If $\beta \neq 0$, then $0^\beta = 0$.
2. $1^\beta = 1$.
3. If $1 < \alpha$ and $\beta < \gamma$, then $\alpha^\beta < \alpha^\gamma$.
4. If $\alpha \leq \beta$, then $\alpha^\gamma \leq \beta^\gamma$.
5. If $1 < \alpha$, then $\beta \leq \alpha^\beta$.

Exercise 3.11 Prove that the following facts about ordinal arithmetic hold for all ordinals α, β and γ.

1. $\alpha^{\beta+\gamma} = \alpha^\beta \cdot \alpha^\gamma$.
2. $(\alpha^\beta)^\gamma = \alpha^{\beta \cdot \gamma}$.

Exercise 3.12 Complete the proof of Lemma 3.37.

Exercise 3.13 Let α and β be ordinals. Prove that if $\beta > 0$, then there are unique ordinals δ and ρ such that $\rho < \beta$ and

$$\alpha = (\beta \cdot \delta) + \rho.$$

Exercise 3.14 Let α be an ordinal such that $\alpha \neq 0$. Prove that there are unique n, β_1, ..., β_n, ℓ_1, ..., ℓ_n such that

- $1 \leq n < \omega$,
- $\alpha \geq \beta_1 > \cdots > \beta_n$,
- $1 \leq \ell_i < \omega$ for every $i = 1, \ldots n$, and
- $\alpha = \omega^{\beta_1} \cdot \ell_1 + \cdots + \omega^{\beta_n} \cdot \ell_n$.

This is called *Cantor normal form*.

Exercise 3.15 For each function $x : \omega \to \omega$, define the *support* of x to be the set

$$\{n < \omega \mid x(n) \neq 0\}.$$

Recall that

$$^{\omega}\omega = \{x \mid x \text{ is a function from } \omega \text{ to } \omega\}.$$

Let

$$A = \{x \in ^{\omega}\omega \mid x \text{ has finite support}\}.$$

Given $x, y \in A$ such that $x \neq y$, there exists a largest $n < \omega$ such that $x(n) \neq y(n)$ and we define

$$x \prec_A y \iff x(n) < y(n).$$

Prove that $(A, \prec_A)$ is a wellordering and

$$\text{type}(A, \prec_A) = \omega^{\omega} \text{ (ordinal exponentiation)}.$$

Exercise 3.16 Find two functions

$$f : \omega \to \omega + \omega$$

and

$$g : \omega + \omega \to \omega + \omega + \omega$$

such that

$$\sup(f[\omega]) = \omega + \omega$$

and

$$\sup(g[\omega + \omega]) = \omega + \omega + \omega$$

but if $h = g \circ f$ is the composition, then

$$\sup(h[\omega]) < \omega + \omega + \omega.$$

Exercise 3.17 Let $\kappa < \lambda < \mu$ be three limit ordinals and

$$f : \kappa \to \lambda$$

and

$$g : \lambda \to \mu$$

be two functions such that

$$\sup(f[\kappa]) = \lambda$$

and

$$\sup(g[\lambda]) = \mu.$$

Assume that g is non-decreasing in the sense that if $\alpha \leq \beta < \lambda$, then $g(\alpha) \leq g(\beta)$. Let $h = g \circ f$ be the composition. Prove that

$$\sup(h[\kappa]) = \mu.$$

4
Cardinality

We now turn from the study of order to that of cardinality, which is a fancy word for size. Cardinal numbers will be defined to be certain kinds of ordinal numbers. Not every ordinal is a cardinal though. The theory here builds on that of the previous chapter.

4.1 Cardinal numbers

Definition 4.1 We say that A *and B have the same cardinality* and write $A \approx B$ iff there is a bijection from A to B.

Granted, it is strange to say that two sets have the same cardinality without having said what *cardinality* means. But we need a lemma before giving that definition.

Lemma 4.2 *For every set A, there exists an ordinal γ such that $\gamma \approx A$.*

Proof The rough idea is to let $f(0)$ be an element of A, then let $f(1)$ be an element of A other than $f(0)$, etc. We keep going until we list all the elements of A as $f(\alpha)$ for some $\alpha < \gamma$. Now we have to make rigorous mathematical sense of this idea. Let

$$\mathcal{F} = \{X \subseteq A \mid X \neq \emptyset\}.$$

By the Axiom of Choice, there exists a choice function $c : \mathcal{F} \to A$ such that, for every $X \in \mathcal{F}$,

$$c(X) \in X.$$

Define $f(\alpha)$ by recursion on ordinals α as follows. If

$$A - f[\alpha] \in \mathcal{F},$$

then let

$$f(\alpha) = c\,(A - f[\alpha]).$$

On the other hand, if

$$A - f[\alpha] \notin \mathcal{F},$$

then leave $f(\beta)$ undefined for every $\beta \geq \alpha$.

Observe that if $\alpha < \beta$ and both $f(\alpha)$ and $f(\beta)$ are defined, then $f(\alpha) \in f[\beta]$ and $f(\beta) \in A - f[\beta]$, hence $f(\alpha) \neq f(\beta)$. This calculation almost shows that f is an injection; what is missing is a proof that f has a set domain.

First suppose there is an ordinal γ such that $f(\gamma)$ is undefined. Let γ be the least such ordinal. Then f is an injection with $\gamma = \operatorname{dom}(f)$ and $\operatorname{ran}(f) \subseteq A$. Also,

$$A - f[\gamma] \notin \mathcal{F},$$

which means exactly that $A - f[\gamma] = \emptyset$. Equivalently, it means that $f[\gamma] = A$. Therefore, f is a bijection from γ to A as desired.

It remains to see that there is an ordinal γ such that $f(\gamma)$ is undefined. Suppose otherwise. Then, for every ordinal α, $f(\alpha)$ is defined. Let

$$S = \{a \in A \mid \text{there exists } \alpha \text{ such that } f(\alpha) = a\}.$$

Then, for every $a \in S$, there exists a unique α such that $f(\alpha) = a$. Apply the Replacement Scheme to this property to conclude that there is a set Ω and a function $g : S \to \Omega$ such that, for every $a \in S$, $f(g(a)) = a$. Because $f(\alpha)$ is defined for every ordinal α, and because $f(\alpha) \neq f(\beta)$ whenever $\alpha \neq \beta$, it must be that Ω is the set of all ordinals. But we proved earlier, in Lemma 3.25, that there is no set of all ordinals. $\qquad\square$

We remark that the use of the Axiom of Choice cannot be eliminated from the proof of Lemma 4.2. Elaborating on the meaning of this remark: If you remove the Axiom of Choice from ZFC the result is known as ZF. It turns out that ZF + Lemma 4.2 implies the Axiom of Choice. Prove this as a practice problem!

At last, we define cardinal numbers and the cardinality of sets.

Definition 4.3 κ is a *cardinal* iff κ is an ordinal and, for every $\eta < \kappa$,

$$\eta \not\approx \kappa$$

Definition 4.4 $|A|$ is the least ordinal κ such that $A \approx \kappa$.

You should convince yourself of the following facts, whose proofs amount to composing bijections.

Lemma 4.5   *$|A|$ is a cardinal.*

Lemma 4.6 $A \approx B \iff |A| = |B|$.

Every natural number is both an ordinal and a cardinal. Also, ω is a cardinal. However, the ordinals

$$\omega + 1, \omega + 2, \omega + 3, \ldots$$

are all *countably infinite*, which is to say that their cardinality is ω. (For us, *countable* means finite or countably infinite. *Uncountable* means not countable.) In particular, the ordinals displayed above are not cardinals. Moreover, the ordinals

$$\omega \cdot 2, \omega \cdot 3, \omega \cdot 4, \ldots$$

are not cardinals. Nor are the ordinals

$$\omega^2, \omega^3, \omega^4, \ldots.$$

The ordinals

$$\omega^\omega, \omega^{\omega^\omega}, \omega^{\omega^{\omega^\omega}} \ldots$$

are not cardinals either. All of these are ordinals are countable, as you should try to verify. We need another idea to reach uncountable sets.

Theorem 4.7 (Cantor)   *There is no surjection from A to $\mathcal{P}(A)$.*

Proof   Consider an arbitrary function $f : A \to \mathcal{P}(A)$. Let

$$C = \{x \in A \mid x \notin f(x)\}.$$

For every $x \in A$,

$$x \in C \iff x \notin f(x)$$

hence

$$C \neq f(x).$$

In particular,

$$C \notin f[A].$$

Therefore $f : A \to \mathcal{P}(A)$ is not a surjection. $\square$

The proof we just gave is an example of a *diagonal argument*. This is an imprecise term that you will see used in more and more general ways throughout the book. Here, the idea is that if you visualize the graph of the relation

$$\{(x, y) \in A \times A \mid x \in f(y)\},$$

then what we call the diagonal is

$$D = \{x \mid x \in f(x)\}.$$

To come up with a set C missing from the range of f, we take the complement of the diagonal,

$$C = A - D.$$

The reason $C \neq f(x)$ is that one of C and $f(x)$ has x as an element, and the other does not. We can express the previous sentence in terms of the symmetric difference:

$$x \in C \triangle f(x)$$

hence

$$C \triangle f(x) \neq \emptyset$$

thus

$$C \neq f(x).$$

Corollary 4.8 $\mathcal{P}(\omega)$ *is uncountable.*

Proof Clearly $\mathcal{P}(\omega)$ is infinite. By Theorem 4.7, there is no surjection from ω to $\mathcal{P}(\omega)$. Hence there is no bijection. $\square$

Bijections are used in the definition of cardinality but sometimes only surjections or injections are easily available. Here are some basic facts that relate these notions.

Lemma 4.9 *If* $\kappa < \theta$ *and there is a surjection* $f : \kappa \to \theta$, *then* θ *is not a cardinal.*

Proof Suppose that $\kappa < \theta$ and $f : \kappa \to \theta$ is a surjection. Let

$$S = \{\beta < \kappa \mid f(\alpha) \neq f(\beta) \text{ for every } \alpha < \beta\}.$$

Let $g = f \restriction S$. Then g is a bijection from S to θ. Let $\sigma = \text{type}(S)$ and $\pi : (S, <) \simeq \sigma$ be the Mostowski collapse of $(S, <)$. By Lemma 3.31, since $S \subseteq \kappa$, $\sigma \leq \kappa$. So $\sigma < \theta$. Let

$$h = g \circ \pi^{-1}.$$

Then $h : \sigma \to \theta$ is a bijection. In other words $\sigma \approx \theta$. Thus θ is not a cardinal. □

Lemma 4.10 *Let κ and λ be cardinals. Then*

$$\kappa \leq \lambda \iff \text{ there is an injection } f : \kappa \to \lambda.$$

Proof If $\kappa \leq \lambda$, then the identity function is an injection from κ to λ. For the reverse direction, suppose that $f : \kappa \to \lambda$ is an injection. Let

$$S = f[\kappa] = \{f(\alpha) \mid \alpha < \kappa\}$$

and $\pi : (S, <) \simeq \sigma$ be the Mostowski collapse. Then the composition

$$\pi \circ f : \kappa \to \sigma$$

is a bijection. Thus $|\kappa| = |\sigma|$. Since κ is a cardinal, $|\kappa| = \kappa$. Because $S \subseteq \lambda$, by Lemma 3.31, $\sigma \leq \lambda$. Putting these facts together we have that

$$\kappa = |\kappa| = |\sigma| \leq \sigma \leq \lambda.$$

□

Corollary 4.11 *For every cardinal κ, there exists a cardinal λ such that $\lambda > \kappa$.*

Proof Let $\lambda = |\mathcal{P}(\kappa)|$. First note that the identity function is an injection from κ to $\mathcal{P}(\kappa)$, which implies that $\kappa \leq \lambda$ by Lemma 4.10. But Theorem 4.7 implies that $\kappa \neq \lambda$, so $\kappa < \lambda$. □

Definition 4.12 κ^+ is the least cardinal strictly greater than κ.

The proof of Corollary 4.11 shows that $\kappa^+ \leq |\mathcal{P}(\kappa)|$. You should not assume that equality holds. The discussion after Corollary 4.27 explains why.

Sometimes, the following theorem is covered in courses other than set theory but with a very different proof. (See Exercise 4.14.) The proof here is extremely short because it builds on the theory of ordinals and cardinals, which we have at hand.

Theorem 4.13 (Cantor–Bernstein–Schroeder) *Suppose that*

- *there is an injection from A to B, and*
- *there is an injection from B to A.*

Then $A \approx B$.

Proof First recall that $A \approx |A|$ and $B \approx |B|$. By Lemma 4.10 and the hypothesis of the theorem, $|A| \leq |B|$ and $|B| \leq |A|$. So $|A| = |B|$. Hence $A \approx B$.　　　　　　　　　　　　　　　　□

By Corollary 4.11, there is no largest cardinal. The next result is a kind of continuity for cardinal numbers.

Lemma 4.14 *If A is a set of cardinals, then $\sup(A)$ is a cardinal.*

Proof We may assume that A does not have a maximum element, as otherwise

$$\sup(A) = \max(A) \in A.$$

For contradiction, suppose that $\sup(A)$ is not a cardinal. Let $\kappa < \sup(A)$ and $f : \kappa \to \sup(A)$ be a surjection. Since A does not have a maximum element, there exists $\lambda \in A$ such that $\kappa < \lambda$. Let

$$S = \{\alpha < \kappa \mid f(\alpha) \in \lambda\}$$

and $g = f \restriction S$. Then $g : S \to \lambda$ is a surjection. Let $\sigma = \text{type}(S)$ and

$$\pi : (S, <) \simeq \sigma$$

be the Mostowski collapse. Because $S \subseteq \kappa$, by Lemma 3.31, $\sigma \leq \kappa$. So $\sigma < \lambda$. Let $h = g \circ \pi^{-1}$. Then $h : \sigma \to \lambda$ is a surjection. Together with Lemma 4.9, this shows that λ is a not a cardinal. But we assumed that every element of A is a cardinal.　　　　　　□

Next we list the infinite cardinals in increasing order using ordinals as indices:

$$\aleph_0, \aleph_1, \aleph_2, \ldots, \aleph_\omega, \aleph_{\omega+1}, \aleph_{\omega+2}, \ldots$$

The letter $\aleph$ is read *aleph* and is the first letter of the Hebrew alphabet. Here is the formal recursive definition of our list of infinite cardinals.

Definition 4.15 Let

$$\aleph_0 = \omega.$$

By recursion on $\beta > 0$, define $\aleph_\beta$ to be the least cardinal greater than $\aleph_\alpha$ for all $\alpha < \beta$.

It is tempting and correct to write

$$\aleph_\beta = \min\left(\left\{\kappa \mid \kappa \text{ is a cardinal and } \kappa > \aleph_\alpha \text{ for all } \alpha < \beta\right\}\right)$$

but keep in mind that what we have inside $\min(\cdot)$ is not a set.

Corollary 4.16 *We have that*

- $\aleph_0 = \omega$,
- $\aleph_{\alpha+1} = (\aleph_\alpha)^+$ *for every ordinal α, and*
- $\aleph_\beta = \sup\left(\{\aleph_\alpha \mid \alpha < \beta\}\right)$ *for every limit ordinal β.*

Proof The first two clauses are obvious. The last clause follows from Lemma 4.14. □

Definition 4.17 We say that λ is a *successor cardinal* iff there is a cardinal κ such that $\lambda = \kappa^+$. If $\lambda \neq 0$ and λ is not a successor cardinal, then we say that λ is a *limit cardinal*.

It is important to note that the only successor ordinals that are cardinals are the natural numbers. Every infinite cardinal (including every infinite successor cardinal) is a limit ordinal. Corollary 4.16 implies the following result, which spells out how these concepts are related.

Corollary 4.18 *For every ordinal α,*

- $\aleph_\alpha$ *is a limit cardinal iff either $\alpha = 0$ or α is a limit ordinal, and*
- $\aleph_\alpha$ *is a successor cardinal iff α is a successor ordinal.*

Corollary 4.18 covers all infinite cardinals by the following fact.

Lemma 4.19 *Let λ be an infinite cardinal. Then there is an ordinal $\beta \leq \lambda$ such that $\lambda = \aleph_\beta$.*

Proof By induction on infinite cardinals λ.

Base case $\lambda = \omega$.

Then $\lambda = \aleph_0$ and $0 < \omega = \lambda$.

Successor case $\lambda = \kappa^+$.

By the induction hypothesis, there is an $\alpha \leq \kappa$ such that $\kappa = \aleph_\alpha$. Then

$$\lambda = \kappa^+ = (\aleph_\alpha)^+ = \aleph_{\alpha+1}$$

by Corollary 4.16, and

$$\alpha + 1 \leq \kappa + 1 < \kappa^+ = \lambda.$$

Limit case λ *is a limit cardinal.*

Let

$$\beta = \sup(\{\alpha \mid \aleph_\alpha < \lambda\}).$$

By the case hypothesis and the induction hypothesis,

$$\begin{aligned}
\beta &\leq \sup(\{\aleph_\alpha \mid \aleph_\alpha < \lambda\}) \\
&= \sup(\{\kappa < \lambda \mid \kappa \text{ is a cardinal}\}) \\
&= \lambda
\end{aligned}$$

and β is a limit ordinal. Similarly,

$$\begin{aligned}
\lambda &= \sup(\{\kappa < \lambda \mid \kappa \text{ is a cardinal}\}) \\
&= \sup(\{\aleph_\alpha \mid \aleph_\alpha < \lambda\}) \\
&= \sup(\{\aleph_\alpha \mid \alpha < \beta\}) \\
&= \aleph_\beta.
\end{aligned}$$

The last line is by Corollary 4.16.

$\square$

4.2 Cardinal arithmetic

Every cardinal is an ordinal but cardinal arithmetic is completely different from ordinal arithmetic when it comes to infinite cardinals. It is important to keep track of which kind of arithmetic you are doing. Usually, it is clear from the context.

Definition 4.20 For all cardinals κ and λ,

$$\kappa \oplus \lambda = |(\{0\} \times \kappa) \cup (\{1\} \times \lambda)|$$

and

$$\kappa \otimes \lambda = |\kappa \times \lambda|.$$

Unlike ordinal addition and multiplication, cardinal addition and multiplication are commutative. The main point in seeing that

$$\kappa \otimes \lambda = \lambda \otimes \kappa$$

is that if A and B are sets, then $(x, y) \mapsto (y, x)$ is a bijection from $A \times B$ to $B \times A$, hence $|A \times B| = |B \times A|$.

The next result says that, for natural numbers, cardinal addition and multiplication coincide with ordinal addition and multiplication. The reader should work out the proofs as an exercise.

Lemma 4.21 *If $m, n < \aleph_0$, then $m \oplus n = m+n$ and $m \otimes n = m \cdot n$.*

Remember that ordinal addition and multiplication for infinite ordinals were interesting operations with subtle properties. By contrast, cardinal addition and multiplication for infinite cardinals turn out to be trivial to calculate by the following two results.

Lemma 4.22 *Let λ be an infinite cardinal. Then $\lambda \otimes \lambda = \lambda$.*

Proof By induction on λ. The induction hypothesis is that

$$\mu \otimes \mu = \mu$$

whenever μ is a cardinal such that $\aleph_0 \leq \mu < \lambda$. We will use two different orderings of the Cartesian product $\lambda \times \lambda$. First define

$$(\overline{\alpha}, \overline{\beta}) <_{\text{lex}} (\alpha, \beta)$$

iff either

$$\overline{\alpha} < \alpha$$

or

$$\overline{\alpha} = \alpha \text{ and } \overline{\beta} < \beta.$$

Then define

$$(\overline{\alpha}, \overline{\beta}) \lhd (\alpha, \beta)$$

iff either

$$\max\left(\overline{\alpha},\overline{\beta}\right) < \max\left(\alpha,\beta\right)$$

or

$$\max\left(\overline{\alpha},\overline{\beta}\right) = \max\left(\alpha,\beta\right) \text{ and } \left(\overline{\alpha},\overline{\beta}\right) <_{\text{lex}} \left(\alpha,\beta\right).$$

Figure 4.1 is a picture of $\lambda \times \lambda$ ordered by $\lhd$. For a given $\alpha < \lambda$, the order $\lhd$ increases across the horizontal arrow (1) leaving out (α, α), then increases up the vertical arrow (2) until it reaches (α, α). Next, $\lhd$ increases across (3) leaving out $(\alpha + 1, \alpha + 1)$, then up (4) including $(\alpha + 1, \alpha + 1)$. And so on.

We claim that $\lhd$ is a wellordering of $\lambda \times \lambda$. It is obviously a strict linear ordering. Towards seeing that $\lhd$ is wellfounded, consider an arbitrary $S \subseteq \lambda \times \lambda$ such that $S \neq \emptyset$. Let

$$\gamma = \min\left(\{\max(\alpha,\beta) \mid (\alpha,\beta) \in S\}\right).$$

Let

$$\overline{\alpha} = \min\left(\{\alpha \mid \text{there is } \beta \text{ with } (\alpha,\beta) \in S \text{ and } \max\left(\alpha,\beta\right) = \gamma\}\right)$$

and

$$\overline{\beta} = \min\left(\{\beta \mid (\overline{\alpha},\beta) \in S\}\right).$$

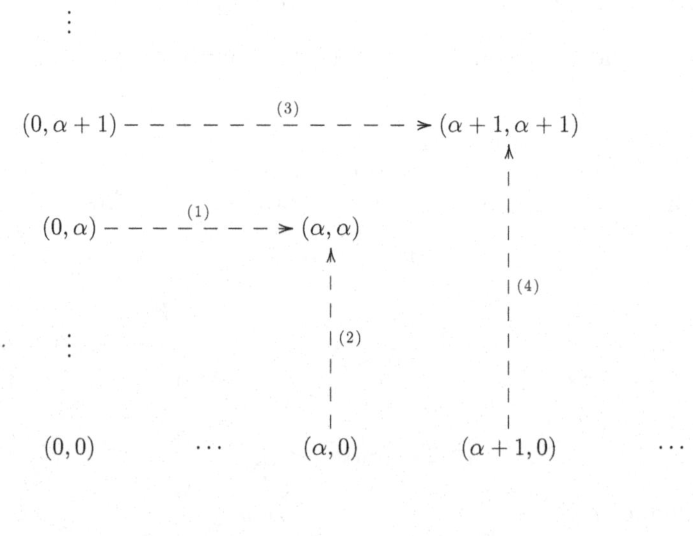

Figure 4.1 $\lambda \times \lambda$ ordered by $\lhd$

It is easy to check that $(\overline{\alpha}, \overline{\beta})$ is the $\lhd$-least element of S. Now that we know $\lhd$ is a wellordering, we can talk about its order type.

Claim 4.22.1 $\text{type} (\lambda \times \lambda, \lhd) = \lambda$.

Proof First note that $\text{type} (\lambda \times \lambda, \lhd) \geq \lambda$. This is because

$$(\overline{\alpha}, 0) \lhd (\alpha, 0)$$

whenever $\overline{\alpha} < \alpha < \lambda$. So it is enough to see that

$$\text{type} (\lambda \times \lambda, \lhd) \leq \lambda.$$

For this, it is enough to see that, for every $(\alpha, \beta) \in \lambda \times \lambda$,

$$\text{type} (\text{pred}_\lhd (\alpha, \beta) , \lhd) < \lambda$$

where, by definition,

$$\text{pred}_\lhd (\alpha, \beta) = \left\{ (\overline{\alpha}, \overline{\beta}) \mid (\overline{\alpha}, \overline{\beta}) \lhd (\alpha, \beta) \right\}.$$

If $\lambda = \aleph_0$, then α and β are natural numbers and $\text{pred}_\lhd (\alpha, \beta)$ is finite, hence

$$\text{type} (\text{pred}_\lhd (\alpha, \beta) , \lhd) < \aleph_0 = \lambda$$

as desired. So we may assume that $\lambda \geq \aleph_1$. Let

$$\gamma = \max(\alpha, \beta, \aleph_0) + 1$$

and

$$\mu = |\gamma|.$$

Then μ is a cardinal and $\aleph_0 \leq \mu < \lambda$, so by the induction hypothesis,

$$\mu \otimes \mu = \mu.$$

Therefore

$$\text{type} (\text{pred}_\lhd (\alpha, \beta) , \lhd) \leq \text{type} (\gamma \times \gamma, \lhd) < \mu^+ \leq \lambda$$

as desired. $\qquad\square$

From Claim 4.22.1, it follows that

$$(\lambda \times \lambda, \lhd) \simeq (\lambda, <).$$

Since isomorphisms are bijections,

$$|\lambda \times \lambda| = |\lambda| = \lambda.$$

This completes the proof of Lemma 4.22. $\qquad\square$

With regard to the proof of Lemma 4.22, we remark that if λ is an ordinal, then the lexigraphic ordering on $\lambda \times \lambda$ is a wellordering of type $\lambda \cdot \lambda$ (ordinal product). In other words,

$$\text{type}(\lambda \times \lambda, <_{\text{lex}}) = \lambda \cdot \lambda.$$

This is by the definition of ordinal multiplication, Definition 3.40. In particular, if λ is an infinite ordinal, then

$$|\lambda| \leq \lambda < \lambda \cdot \lambda < |\lambda|^+.$$

In particular, $\lambda \cdot \lambda$ is not a cardinal.

Lemma 4.22 is a special case of the following general result.

Theorem 4.23 *If $0 < \kappa \leq \lambda$ are cardinals and $\aleph_0 \leq \lambda$, then*

$$\kappa \oplus \lambda = \kappa \otimes \lambda = \lambda.$$

Proof The theorem can be verified easily if $\kappa = 1$. If $\kappa \geq 2$, then by Lemma 4.22,

$$\lambda \leq \kappa \oplus \lambda \leq \lambda \oplus \lambda = 2 \otimes \lambda \leq \kappa \otimes \lambda \leq \lambda \otimes \lambda = \lambda$$

so equality holds throughout. $\qquad\qquad\qquad\qquad\qquad\qquad\qquad$ $\square$

Theorem 4.23 tells us that if at least one of κ and λ is infinite, then

$$\kappa \oplus \lambda = \kappa \otimes \lambda = \max(\kappa, \lambda).$$

So cardinal addition and multiplication really are easy to calculate!

Now we define cardinal exponentiation. The notation is the same as for ordinal exponentiation but the meaning is different. For this definition, recall from Exercise 2.8 that if A and B are sets, then

$$^{A}B = \{f \mid f \text{ is a function from } A \text{ to } B\}.$$

Definition 4.24 For all cardinals κ and λ,

$$\lambda^{\kappa} = |^{\kappa}\lambda|$$

We repeat our warning that writing λ^{κ} is ambiguous. Do you mean ordinal exponentiation or cardinal exponentiation? Always make sure it is clear which, either from the context, or by saying so explicitly.

Our first fact about cardinal exponentiation is that it is the same as ordinal exponentiation when restricted to the natural numbers. The reader should work out the proof as an exercise.

Lemma 4.25 *If $m, n < \aleph_0$, then n^m is the same whether computed as ordinal exponentiation or cardinal exponentiation.*

Cardinal exponentiation becomes quite interesting when we look at infinite powers.

Lemma 4.26 $|\mathcal{P}(A)| = 2^{|A|}$

Proof Define a function

$$\text{char} : \mathcal{P}(A) \to {}^A 2$$

by setting

$$\text{char}(X)(a) = \begin{cases} 0 & \text{if } a \notin X \\ 1 & \text{if } a \in X \end{cases}$$

for every $X \subseteq A$ and $a \in A$. Note that char is a function whose outputs are themselves functions. The function

$$\text{char}(X) : A \to 2$$

is called the *characteristic function* of X in A.[1] To see that char is an injection, note that if $X, Y \subseteq A$ and $X \neq Y$, then, for any $a \in X \triangle Y$,

$$\text{char}(X)(a) \neq \text{char}(Y)(a),$$

so

$$\text{char}(X) \neq \text{char}(Y).$$

To see that char is a surjection, note that if $f \in {}^A 2$, then

$$\text{char}(\{a \in A \mid f(a) = 1\}) = f.$$

We have shown that $\mathcal{P}(A) \approx {}^A 2$. From this, Lemma 4.26 is clear. ∎

Earlier, we established that $\kappa^+ \leq |\mathcal{P}(\kappa)|$. Thus, the following is a consequence of Lemma 4.26.

Corollary 4.27 *If κ is a cardinal, then $2^\kappa \geq \kappa^+$.*

---

[1] Elsewhere, the characteristic function of X is written χ_X, where χ is the lower case Greek letter chi, and X is a subscript. But this is hard to read.

This brings up an important question, namely, what is the value of 2^κ? Focusing on the most basic case, what is the value of $2^{\aleph_0}$? Is $2^{\aleph_0} = \aleph_1$? Perhaps $2^{\aleph_0} = \aleph_2$? Could it be that $2^{\aleph_0} = \aleph_{\omega \cdot 7 + 4}$? We know that, for some ordinal $\alpha \geq 1$,

$$2^{\aleph_0} = \aleph_\alpha$$

but it turns out that the value of α cannot be determined using only the axioms of ZFC because of deep theorems of Kurt Gödel and Paul Cohen.[2] This is interesting because

$$|\mathbb{R}| = 2^{\aleph_0},$$

so really we are asking how many real numbers there are. This problem was posed by Georg Cantor in the late 1800s, who asked:

Is there is an uncountable $A \subseteq \mathbb{R}$ such that $A \not\approx \mathbb{R}$?

It was also first on the most famous list of open problems, which David Hilbert compiled at the start of the twentieth century. The answer *no* to Cantor's question is known as the Continuum Hypothesis, or CH, which says $2^{\aleph_0} = \aleph_1$. The answer *yes* says that $2^{\aleph_0} \geq \aleph_2$. Cantor conjectured CH is true, but an informal poll suggests that most set theorists today who have an opinion believe CH is counterintuitive. Few have strong feelings about what the actual value of $2^{\aleph_0}$ should be although some feel it should be $\aleph_2$. As we mentioned already, there are theorems due to Gödel and Cohen which together say roughly that ZFC is not powerful enough to answer Cantor's question, so it is unknown what methodology could lead to an answer. Additional explanation would be beyond the scope of this book; it should be the topic of your next set theory course!

Remember that, for finite numbers, $(m^\ell)^k = m^{\ell k}$. Do not confuse this with the standard convention $m^{\ell^k} = m^{(\ell^k)}$. The first equation generalizes to all cardinal numbers as the following lemma shows. Other basic facts about cardinals can be found in the exercises.

---

[2] The hypothesis of these Gödel and Cohen theorems is that ZFC is a consistent theory, meaning there is no proof that $0 = 1$ using only the axioms of ZFC. Gödel proved that ZFC is consistent with CH in 1940. Cohen proved that ZFC is consistent with the negation of CH in 1963. The combination of these results says CH is *independent* of ZFC. These results can be found in many graduate level textbooks on set theory.

Lemma 4.28 *Let κ, λ and μ be cardinals. Then*

$$(\mu^\lambda)^\kappa = \mu^{\lambda \otimes \kappa}.$$

Proof It is easy to see that

$$\left(\mu^\lambda\right)^\kappa = \left|{}^\kappa\left({}^\lambda\mu\right)\right|$$

and

$$\mu^{\lambda \otimes \kappa} = \left|{}^{\kappa \times \lambda}\mu\right|.$$

There is a bijection

$$F : {}^\kappa\left({}^\lambda\mu\right) \to {}^{\kappa \times \lambda}\mu$$

defined by

$$F(g)(\alpha, \beta) = g(\alpha)(\beta).$$

Putting together these observations, we are done. $\square$

Definition 4.29 (Strange notation) $\omega_\alpha = \aleph_\alpha$ for ordinals $\alpha \geq 1$.

We write ω_α when we want to emphasize that it is an ordinal. We write $\aleph_\alpha$ when we want to emphasize that it is a cardinal. However, it is difficult to keep the notation consistent with such intentions when we simultaneously consider cardinal and ordinal properties of $\aleph_\alpha = \omega_\alpha$. Ultimately, whether we write ω_α or $\aleph_\alpha$ reduces to a matter of style.

Note that ω_0 is not defined; we always write either ω or $\aleph_0$.

4.3 Cofinality

Recall that $\aleph_\omega$ is the least cardinal greater than every $\aleph_n$ for $n < \omega$. In this sense, $\aleph_\omega$ feels rather large. On the other hand, the function

$$n \mapsto \aleph_n$$

maps ω to $\aleph_\omega$ and has range unbounded in $\aleph_\omega$, i.e.,

$$\aleph_\omega = \sup_{n<\omega} \aleph_n.$$

Let us point out that $\omega = \aleph_0 < \aleph_\omega$. The fact that $\aleph_\omega$ can be reached from below in this way makes it feel somewhat smaller than before. Here, we examine this phenomenon and tie it up with cardinal arithmetic.

Definition 4.30   If λ is a limit ordinal, then the *cofinality of λ*,

$$\mathrm{cf}(\lambda),$$

is the least ordinal κ such that there exists a function $f : \kappa \to \lambda$ with

$$\sup(f[\kappa]) = \lambda.$$

We say λ is *singular* if $\mathrm{cf}(\lambda) < \lambda$. Otherwise, we say λ is *regular*.

The notions above are defined only for limit ordinals, not for 0 or successor ordinals $\alpha + 1$. Keep in mind that every infinite cardinal is a limit ordinal. (Recall the reason is that $|A \cup \{A\}| = |A|$ for every infinite set A.) Here is a list of examples of cofinalities to think about now and as you read.

$$\mathrm{cf}\,(\aleph_0) = \aleph_0$$
$$\mathrm{cf}\,(\aleph_1) = \aleph_1$$
$$\mathrm{cf}\,(\aleph_2) = \aleph_2$$
$$\mathrm{cf}\,(\aleph_\omega) = \aleph_0$$
$$\mathrm{cf}\,(\aleph_{\omega+1}) = \aleph_{\omega+1}$$
$$\mathrm{cf}\,(\aleph_{\omega+2}) = \aleph_{\omega+2}$$
$$\mathrm{cf}\,(\aleph_{\omega+\omega}) = \aleph_0$$
$$\mathrm{cf}\,(\aleph_{\omega_1}) = \aleph_1$$

Here is a second list with more examples to think about.

$$\mathrm{cf}\,(\omega + \omega) = \aleph_0$$
$$\mathrm{cf}\,(\omega \cdot \omega) = \aleph_0$$
$$\mathrm{cf}\,(\omega^\omega) = \aleph_0 \qquad \text{(ordinal exponentiation)}$$
$$\mathrm{cf}\,(\omega_1 + \omega) = \aleph_0$$
$$\mathrm{cf}\,(\omega_1 + \omega_1) = \aleph_1$$
$$\mathrm{cf}\,(\omega_1 + \omega_1 + \omega) = \aleph_0$$

By the end of this chapter, you should be able to explain the equations listed above. Currently, you should see that, for every limit ordinal λ,

$$\mathrm{cf}(\lambda) \leq |\lambda| \leq \lambda.$$

This is because every surjection onto λ has range unbounded in λ. It follows from these inequalities that if $\mathrm{cf}(\lambda) = \lambda$, then $|\lambda| = \lambda$. In other words, if λ is a regular limit ordinal, then λ is a cardinal.

It is worth doing Exercises 3.16 and 3.17 before reading the proof of the next result.

Lemma 4.31 $\mathrm{cf}(\lambda)$ *is a regular cardinal.*

Proof Let $f : \mathrm{cf}(\lambda) \to \lambda$ be a function whose range is unbounded in λ. We know there is such a function f by the definition of cofinality.

The following claim records slightly more information than we need for the rest of the proof but the extra information is useful elsewhere.

Claim 4.31.1 *There is a function* $g : \mathrm{cf}(\lambda) \to \lambda$ *such that*

- $\mathrm{ran}(g)$ *is unbounded in* λ,
- g *is non-decreasing in the sense that if* $\alpha \leq \beta < \mathrm{cf}(\lambda)$, *then* $g(\alpha) \leq g(\beta)$,
- g *is continuous in the sense that for every limit ordinal* $\beta < \mathrm{cf}(\lambda)$,

$$g(\beta) = \sup_{\alpha < \beta} g(\alpha).$$

Proof of claim Define g with $\mathrm{dom}(g) = \mathrm{cf}(\lambda)$ by

$$g(\beta) = \sup_{\alpha < \beta} f(\alpha).$$

By the definition of cofinality, if $\beta < \mathrm{cf}(\lambda)$, then $g(\beta) < \lambda$. That is,

$$g : \mathrm{cf}(\lambda) \to \lambda.$$

To see that g is continuous, observe that, for every limit ordinal $\gamma < \mathrm{cf}(\lambda)$,

$$g(\gamma) = \sup_{\alpha < \gamma} f(\alpha) = \sup_{\beta < \gamma} \left(\sup_{\alpha < \beta} f(\alpha) \right) = \sup_{\beta < \gamma} g(\beta).$$

Clearly, $\operatorname{cf}(\lambda)$ is a limit ordinal. Using this, we see that the range of g is unbounded in λ because

$$\sup_{\beta < \operatorname{cf}(\lambda)} g(\beta) = \sup_{\beta < \operatorname{cf}(\lambda)} \left(\sup_{\alpha < \beta} f(\alpha) \right) = \sup_{\alpha < \operatorname{cf}(\lambda)} f(\alpha) = \lambda.$$

$\square$

Continuing with the proof of Lemma 4.31, let $g : \operatorname{cf}(\lambda) \to \lambda$ be as in Claim 4.31.1. Since every regular limit ordinal is a cardinal, in order to finish proving the lemma, it suffices to show that $\operatorname{cf}(\lambda)$ is a regular ordinal. That is, given an ordinal $\kappa < \operatorname{cf}(\lambda)$ and a function $h : \kappa \to \operatorname{cf}(\lambda)$, we must conclude that the range of h is bounded in $\operatorname{cf}(\lambda)$. Let $S = h[\kappa]$. For contradiction, assume that

$$\sup(S) = \operatorname{cf}(\lambda).$$

Using this assumption and the fact that g is non-decreasing and unbounded, we see that

$$\sup_{\eta < \kappa} g(h(\eta)) = \sup_{\alpha \in S} g(\alpha) = \sup_{\alpha < \operatorname{cf}(\lambda)} g(\alpha) = \lambda.$$

We have shown that the composition

$$g \circ h : \kappa \to \lambda$$

has range unbounded in λ. Because $\kappa < \operatorname{cf}(\lambda)$, this contradicts the definition of cofinality. $\square$

The following result implies that $\operatorname{cf}(\aleph_{\alpha+1}) = \aleph_{\alpha+1}$ for every ordinal α.

Lemma 4.32 *Let λ be an infinite cardinal and $\mu = \lambda^+$. Then μ is a regular cardinal.*

Proof We must show that $\operatorname{cf}(\mu) = \mu$. Let $\kappa \le \lambda$ and $f : \kappa \to \mu$ be a function. For each $\alpha < \kappa$, $f(\alpha) < \mu = \lambda^+$, so $|f(\alpha)| \le \lambda$. Thus

$$|\sup(f[\kappa])| \le \left| \bigcup_{\alpha < \kappa} f(\alpha) \right| \le \lambda \otimes \kappa = \lambda < \mu.$$

In particular, $\sup(f[\kappa]) < \mu$. This shows that $\operatorname{cf}(\mu) \ge \mu$. But obviously $\operatorname{cf}(\mu) \le \mu$ for every limit ordinal μ. Hence $\operatorname{cf}(\mu) = \mu$, which means μ is regular. $\square$

Now that we understand successor cardinals, let us look at a few examples of singular cardinals.

Example Using the fact that

$$\aleph_\omega = \sup_{n<\omega} \aleph_n$$

we see that

$$\mathrm{cf}(\aleph_\omega) = \aleph_0.$$

Example Using the fact that

$$\aleph_{\omega+\omega} = \sup_{n<\omega} \aleph_{\omega+n},$$

we see that

$$\mathrm{cf}(\aleph_{\omega+\omega}) = \aleph_0.$$

Example Using the fact that

$$\aleph_{\omega_1} = \sup_{\alpha<\omega_1} \aleph_\alpha,$$

we see that

$$\mathrm{cf}(\aleph_{\omega_1}) \leq \aleph_1.$$

We claim that

$$\mathrm{cf}(\aleph_{\omega_1}) \neq \aleph_0.$$

Consider an arbitrary function

$$f : \omega \to \aleph_{\omega_1}.$$

Define $g : \omega \to \omega_1$ by letting

$$g(n) = \text{the least } \alpha < \omega_1 \text{ such that } f(n) < \aleph_\alpha.$$

Then, for every $n < \omega$,

$$|f(n)| \leq f(n) < \aleph_{g(n)} < \aleph_{\omega_1}.$$

Since $\aleph_1$ is a regular cardinal, there exists $\alpha < \omega_1$ such that, for every $n < \omega$, $g(n) < \alpha$. Hence $f(n) < \aleph_\alpha$ for every $n < \omega$. So

$$\sup_{n<\omega} f(n) \leq \aleph_\alpha < \aleph_{\omega_1}.$$

This proves our claim. We conclude that

$$\mathrm{cf}(\aleph_{\omega_1}) = \aleph_1.$$

The solution to Exercise 4.13 involves calculations similar to those in our examples above.

The following theorem is an extension of Theorem 4.7. Its proof is a more elaborate diagonal argument.

Theorem 4.33 (G. Kőnig) *If λ is an infinite cardinal, then*

$$\lambda^{\mathrm{cf}(\lambda)} > \lambda.$$

Proof Let $\kappa = \mathrm{cf}(\lambda)$. Fix $f : \kappa \to \lambda$ such that the range of f is unbounded in λ. Consider an arbitrary function $G : \lambda \to {}^\kappa\lambda$. It is enough to see that G is not a surjection. For each $\alpha < \kappa$, let

$$A_\alpha = \{G(\eta)(\alpha) \mid \eta < f(\alpha)\}.$$

Then, for every $\alpha < \kappa$,

$$A_\alpha \subseteq \lambda$$

and

$$|A_\alpha| \leq f(\alpha) < \lambda.$$

In particular, for every $\alpha < \kappa$,

$$\lambda - A_\alpha \neq \emptyset.$$

For each $\alpha < \kappa$, let $h(\alpha)$ be the least element of $\lambda - A_\alpha$. Then, for every $\alpha < \kappa$ and every $\eta < f(\alpha)$,

$$h(\alpha) \neq G(\eta)(\alpha).$$

Recall that, for every $\eta < \lambda$, there exists $\alpha < \kappa$ such that $\eta < f(\alpha)$. Therefore, for every $\eta < \lambda$,

$$h \neq G(\eta).$$

This shows that G is not a surjection. □

Corollary 4.34 *If κ is an infinite cardinal, then*

$$\mathrm{cf}(2^\kappa) > \kappa.$$

Proof Apply the previous theorem with $\lambda = 2^\kappa$ to see that

$$(2^\kappa)^{\mathrm{cf}(2^\kappa)} > 2^\kappa.$$

But, if $\mu \leq \kappa$, then

$$(2^\kappa)^\mu = 2^{\kappa \otimes \mu} = 2^\kappa$$

by Lemmas 4.28 and Theorem 4.23. Corollary 4.34 follows. □

An interesting special case of Corollary 4.34 is the fact that

$$\text{cf}(2^{\aleph_0}) > \aleph_0.$$

Notice, also, that we recover Theorem 4.7 from Corollary 4.34 because

$$2^{\kappa} \geq \text{cf}\,(2^{\kappa}) > \kappa.$$

Exercises

Exercise 4.1 Let

$$^{<\omega}2 = \bigcup_{n<\omega} {}^{n}2.$$

1. Prove that $^{<\omega}2$ is countable.
2. Let

$$\mathcal{F} = \{\{x \upharpoonright n \mid n < \omega\} \mid x \in {}^{\omega}2\}.$$

   Prove that $|\mathcal{F}| = 2^{\aleph_0}$.
3. Prove that there exists a family $\mathcal{G} \subseteq \mathcal{P}(\omega)$ such that

$$|\mathcal{G}| = 2^{\aleph_0}$$

   and for all $A, B \in \mathcal{G}$, if $A \neq B$, then $A \cap B$ is finite.
   Hint: Observe that $\mathcal{F} \subseteq \mathcal{P}(^{<\omega}2)$.

Exercise 4.2 Let

$$^{<\omega}\omega = \bigcup_{n<\omega} {}^{n}\omega$$

and

$$\aleph_0^{<\aleph_0} = \left|^{<\omega}\omega\right|.$$

Show that $\aleph_0^{<\aleph_0} = \aleph_0$.

Exercise 4.3 Prove the following equations.

1. $\aleph_0^{\aleph_0} = 2^{\aleph_0}$
2. $\aleph_1^{\aleph_0} = 2^{\aleph_0}$
3. $\aleph_0^{\aleph_1} = 2^{\aleph_1}$
4. $\aleph_1^{\aleph_1} = 2^{\aleph_1}$

Exercise 4.4   Let

$$^{<\kappa}\lambda = \bigcup_{\alpha<\kappa} {}^\alpha\lambda$$

and

$$\lambda^{<\kappa} = \left|{}^{<\kappa}\lambda\right|$$

whenever κ and λ are infinite cardinals. Show that $\kappa^{(<\kappa^+)} = 2^\kappa$.

Exercise 4.5   Prove that if $\kappa \leq \lambda$ are infinite cardinals, then

$$|\{X \subseteq \lambda \mid |X| = \kappa\}| = \lambda^\kappa.$$

Then explain why

$$|\{X \subseteq \omega_2 \mid |X| = \aleph_0\}| = \max(\aleph_2, 2^{\aleph_0}),$$

$$|\{X \subseteq \omega_2 \mid |X| = \aleph_1\}| = 2^{\aleph_1}$$

and

$$|\{X \subseteq \omega_2 \mid |X| = \aleph_2\}| = 2^{\aleph_2}.$$

Exercise 4.6   Prove that, for every ordinal ξ, there is a cardinal $\lambda > \xi$ such that

$$\mathrm{cf}(\lambda) = \omega$$

and

$$\lambda = \aleph_\lambda.$$

Hint: First recall that $\alpha \leq \aleph_\alpha$ for every ordinal α by Lemma 4.19. Now consider the sequence of cardinals $\langle \kappa_n \mid n < \omega \rangle$ defined by induction according to $\kappa_0 = \aleph_{\xi+1}$ and $\kappa_{n+1} = \aleph_{\kappa_n}$.

Exercise 4.7   Express the cardinality of the sets below in the form

$$\aleph_\alpha, \ 2^{\aleph_\alpha}, \ 2^{2^{\aleph_\alpha}}, \ \ldots$$

and explain your calculations. For your solutions, you may use facts about $\mathbb{Q}, \mathbb{R}, \mathbb{C}$ and continuous functions from calculus courses.

1. $\mathbb{Q} = \{x \mid x \text{ is a rational number}\}$
2. $\mathbb{R} = \{x \mid x \text{ is a real number}\}$
3. $\mathbb{R} - \mathbb{Q} = \{x \in \mathbb{R} \mid x \text{ is irrational}\}$
4. $\{x \in \mathbb{R} \mid 0 < x < 1\}$
5. $\{x \in \mathbb{C} \mid x \text{ is a root of a polynomial with rational coefficients}\}$

6. $^{\mathbb{R}}\mathbb{R} = \{f \mid f \text{ is a function from } \mathbb{R} \text{ to } \mathbb{R}\}$
7. $^{\mathbb{Q}}\mathbb{R} = \{f \mid f \text{ is a function from } \mathbb{Q} \text{ to } \mathbb{R}\}$
8. $^{\mathbb{R}}\mathbb{Q} = \{f \mid f \text{ is a function from } \mathbb{R} \text{ to } \mathbb{Q}\}$
9. $\{f \mid f \text{ is a continuous function from } \mathbb{R} \text{ to } \mathbb{R}\}$

Exercise 4.8   Let $<_{\mathbb{R}}$ be the usual ordering of $\mathbb{R}$. For every $A \subseteq \mathbb{R}$, we may abuse notation by writing

$$(A, <_{\mathbb{R}})$$

when we really mean

$$(A, \{(x, y) \in A \times A \mid x <_{\mathbb{R}} y\}).$$

You may use facts from calculus courses in your solutions to:

1. Prove that, for every $A \subseteq \mathbb{R}$, if $(A, <_{\mathbb{R}})$ is a wellordering, then

$$\text{type}(A, <_{\mathbb{R}}) < \omega_1.$$

2. Prove that, for every $\alpha < \omega_1$, there exists $A_\alpha \subseteq \mathbb{R}$ such that

$$\text{type}(A_\alpha, <_{\mathbb{R}}) = \alpha.$$

Exercise 4.9   Let $f : B \to B$ be a function and $X \subseteq B$. Prove that there exists $A \subseteq B$ such that $X \subseteq A$, $f[A] \subseteq A$ and

$$|A| \leq |X| \otimes \aleph_0.$$

Exercise 4.10   If $\langle S_n \mid n < \omega \rangle$ is a sequence of sets, then define

$$\prod_{n<\omega} S_n = \left\{ f \;\middle|\; \begin{array}{l} f \text{ is a function with } \text{dom}(f) = \omega \\ \text{and } f(n) \in S_n \text{ for all } n < \omega \end{array} \right\}.$$

1. Show that

$$|\mathcal{P}(\aleph_\omega)| \geq \left| \prod_{n<\omega} \mathcal{P}(\aleph_n) \right|.$$

2. Show that

$$|\mathcal{P}(\aleph_\omega)| \leq \left| \prod_{n<\omega} \mathcal{P}(\aleph_n) \right|.$$

Hint: Consider the function

$$A \mapsto \langle A \cap \aleph_n \mid n < \omega \rangle$$

We remark that, from Exercise 4.10, it is immediate that

$$2^{\aleph_\omega} = \left| \prod_{n<\omega} 2^{\aleph_n} \right|,$$

which is usually abbreviated

$$2^{\aleph_\omega} = \prod_{n<\omega} 2^{\aleph_n}.$$

Exercise 4.11   Prove that, for every $n < \omega$,

$$(\aleph_n)^{\aleph_0} = \max(\aleph_n, (\aleph_0)^{\aleph_0}).$$

Hint: It is obvious that the left side is at least as large as the right side. To prove the other direction, use induction on $n < \omega$ and the fact that $\aleph_n$ is a regular cardinal.

Exercise 4.12   This exercise is about ordinal exponentiation.

1. Show that if $\alpha < \omega_1$, then $\omega^\alpha < \omega_1$.
2. Show that $\omega^{\omega_1} = \omega_1$.
3. Show that $\{\beta < \omega_1 \mid \omega^\beta = \beta\}$ is uncountable.

Exercise 4.13   Prove that, for every limit ordinal α,

$$\mathrm{cf}(\aleph_\alpha) = \mathrm{cf}(\alpha).$$

Exercise 4.14 (Cantor–Bernstein–Schroeder theorem)   The proof we gave of Theorem 4.13 used the Axiom of Choice because it used the fact that every set has a cardinality. Complete the following outline of a proof that avoids the Axiom of Choice. Let

$$f : A \to B$$

and

$$g : B \to A$$

be injections. By recursion, define $A_0 = A$, $B_0 = B$,

$$A_{n+1} = g[B_n]$$

and

$$B_{n+1} = f[A_n].$$

Let

$$A_{\text{even}} = \bigcup_{n<\omega} (A_{2n} - A_{2n+1}),$$

$$A_{\text{odd}} = \bigcup_{n<\omega} (A_{2n+1} - A_{2n+2})$$

and

$$A_{\infty} = \bigcap_{n<\omega} A_n.$$

Define

$$h(x) = \begin{cases} f(x) & \text{if } x \in A_{\text{even}} \\ g^{-1}(x) & \text{if } x \in A_{\text{odd}} \\ f(x) & \text{if } x \in A_{\infty}. \end{cases}$$

Prove that h is well-defined and h is a bijection from A to B.

Exercise 4.15 As in Exercises 2.10 and 2.13, let E be the equivalence relation on $\mathcal{P}(\omega)$ defined by

$$xEy \iff x \bigtriangleup y \text{ is finite.}$$

Prove that $\mathcal{P}(\omega)/E$ has cardinality $2^{\aleph_0}$.

Exercise 4.16 (Zorn's lemma) Let $(P, \trianglelefteq)$ be a *partial ordering*. By definition this means that $\trianglelefteq$ is a relation on P that is reflexive and transitive. A subset $C \subseteq P$ is called a *chain* iff $(C, \trianglelefteq)$ is a linear ordering. Since we already know that $\trianglelefteq$ is reflexive and transitive, if $C \subseteq P$, then C is a chain iff for all $x, y \in C$, either $x \trianglelefteq y$ or $y \trianglelefteq x$. Assume that every chain has an *upper bound* in $(P, \trianglelefteq)$. In other words, assume that, for every chain C, there exists $y \in P$ such that, for every $x \in C$, $x \trianglelefteq y$. Prove that $(P, \trianglelefteq)$ has a *maximal element*. In other words, prove that there exists $y \in P$ such that, for every $x \in P$, $y \ntrianglelefteq x$.

Hint: Suppose otherwise. Let $\kappa = |P|$. By recursion on $\alpha < \kappa$, build a chain $C = \{x_\alpha \mid \alpha < \kappa\}$ such that C does not have an upper bound to get a contradiction.

Remark: This proof of Zorn's lemma uses the Axiom of Choice to know that the partial ordering has a cardinality. It also turns out that ZF together with Zorn's lemma implies AC. Therefore, Zorn's lemma and AC are equivalent. This is another good exercise!

Exercise 4.17 (Boolean algebras of truth tables) For each positive $n < \omega$, define

$$\mathbb{T}_n = (T_n, \vee_n, \wedge_n, \neg_n, \overline{0}, \overline{1})$$

as follows.

- T_n is the set of all functions f from $^{\omega}2$ to 2 with the property that, for all $a, b \in {}^{\omega}2$, if $a \restriction n = b \restriction n$, then $f(a) = f(b)$.
- If $f, g \in T_n$ and $a \in {}^{\omega}2$, then

$$(f \vee_n g)(a) = 1 \iff f(a) = 1 \text{ or } g(a) = 1,$$

$$(f \wedge_n g)(a) = 1 \iff f(a) = 1 \text{ and } g(a) = 1$$

and

$$(\neg_n f)(a) = 1 \iff f(a) = 0.$$

- For every $a \in {}^{\omega}2$, $\overline{0}(a) = 0$ and $\overline{1}(a) = 1$.

Notice that $\overline{0}, \overline{1} \in T_n$ for every $n < \omega$.

You may take it for granted that each $\mathbb{T}_n$ is a Boolean algebra. It is helpful to think of $\mathbb{T}_n$ as the Boolean algebra of truth tables in n variables. For example, a typical element f of T_2 can be thought of as the truth table

0	0	$f(\langle 0, 0, \dots \rangle)$
0	1	$f(\langle 0, 1, \dots \rangle)$
1	0	$f(\langle 1, 0, \dots \rangle)$
1	1	$f(\langle 1, 1, \dots \rangle)$

and, if f happens to be an element of T_1, then f can be thought of as the simpler truth table

0	$f(\langle 0, \dots \rangle)$
1	$f(\langle 1, \dots \rangle)$

1. How many elements does T_n have? Explain.
2. Find a finite Boolean algebra $\mathbb{B}$ such that, for every $n < \omega$,

$$\mathbb{B} \ncong \mathbb{T}_n.$$

3. The Boolean algebra relation for $\mathbb{T}_n$ is given by

$$f \preccurlyeq_n g \iff f \wedge_n g = f.$$

   Give a more practical description of $\preccurlyeq_n$ in terms of entries in truth tables.

4. Figure 4.2 shows the elements of T_2 organized into levels with some arrows drawn between some truth tables. What is the significance of the arrows? Which arrows are missing? Copy the figure and add all the missing arrows between truth tables on neighboring levels.

5. List all the atoms of $\mathbb{T}_1$ using truth table notation. Where are they on Figure 4.2? List all the atoms of $\mathbb{T}_2$. Where are they on Figure 4.2? How many atoms does $\mathbb{T}_3$ have?

6. Define

$$\mathbb{T}_\infty = (T_\infty, \vee_\infty, \wedge_\infty, \neg_\infty, \bar{0}, \bar{1})$$

   by setting

$$T_\infty = \bigcup_{n<\omega} T_n,$$

$$\vee_\infty = \bigcup_{n<\omega} \vee_n,$$

$$\wedge_\infty = \bigcup_{n<\omega} \wedge_n$$

   and

$$\neg_\infty = \bigcup_{n<\omega} \neg_n.$$

To make sure you understand the definition of $\mathbb{T}_\infty$, convince yourself that if $f \in T_m$ and $g \in T_n$ where $m < n < \omega$, then $f \in T_n$ and $f \vee_\infty g = f \vee_n g$.

(a) It is a fact that $\mathbb{T}_\infty$ is a Boolean algebra. Pick any three of the ten defining equations for Boolean algebras and show that they hold for $\mathbb{T}_\infty$. You may use the fact that $\mathbb{T}_n$ is a Boolean algebra for $n < \omega$.

(b) Prove that $\mathbb{T}_\infty$ has no atoms.

(c) Explain why T_∞ is countable.

(d) Give a specific example of a function $f : \omega \to 2$ that does not belong to T_∞.

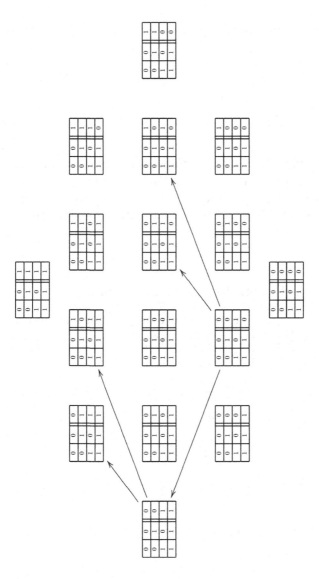

Figure 4.2 *What is the significance of this picture? See Exercise 4.17.*

5

Trees

As you might expect, trees play important roles in many parts of mathematics. Most of this chapter is concerned with trees of height at most ω but the last section goes into trees of height ω_1. We will look at trees in various contexts: topology, analysis, combinatorics and games.

5.1 Topology fundamentals

To get started, we go over some elementary definitions and facts about topological spaces and metric spaces.

Definition 5.1 A *topological space* is a pair $(S, \mathcal{T})$ such that

- $\mathcal{T} \subseteq \mathcal{P}(S)$,
- $S \in \mathcal{T}$,
- for every non-empty finite $\mathcal{F} \subseteq \mathcal{T}$,

$$\bigcap \mathcal{F} \in \mathcal{T},$$

- for every $\mathcal{F} \subseteq \mathcal{T}$,

$$\bigcup \mathcal{F} \in \mathcal{T}.$$

We also say that $\mathcal{T}$ *is a topology on* S.

The most important example of a topological space $(S, \mathcal{T})$ has

$$S = \mathbb{R}$$

and

$$\mathcal{T} = \{U \mid U \text{ is an open subset of } \mathbb{R}\}$$

where U is an open subset of $\mathbb{R}$ iff U is a union of open intervals. For the record, open intervals of $\mathbb{R}$ are sets of the form

$$\{x \in \mathbb{R} \mid a < x < b\}$$

where $a < b$ are real numbers ordered in the usual way. In this section, we will use the notation (a, b) for the open interval from a to b even though it conflicts with our notation for ordered pairs. Other basic notation from calculus may also be used here. Some examples of open subsets of $\mathbb{R}$ are

$$\mathbb{R} = \bigcup \{(-n, n) \mid n = 1, 2, 3, \dots\},$$

$$\emptyset = \bigcup \emptyset,$$

$$(0, 1) = \bigcup \{(0, 1)\},$$

$$(0, \infty) = \bigcup \{(0, n) \mid n = 1, 2, 3, \dots\},$$

$$(-\infty, 0) = \bigcup \{(-n, 0) \mid n = 1, 2, 3, \dots\},$$

$$\mathbb{R} - \{0\} = (-\infty, 0) \cup (0, \infty)$$

and

$$\mathbb{R} - \mathbb{Z} = \bigcup \{(n, n+1) \mid n \in \mathbb{Z}\}.$$

The following fact is left as an exercise; we will give a similar proof in the next section.

Lemma 5.2   *The family of open subsets of $\mathbb{R}$ is a topology on $\mathbb{R}$.*

Topological spaces are related to metric spaces, which we define next.

Definition 5.3   A *metric space* is a pair (S, d) where

$$d : S \times S \to [0, \infty) = \{x \in \mathbb{R} \mid 0 \le x\}$$

is a function from $S \times S$ to the set of non-negative real numbers such that, for all $x, y, z \in S$,

$$d(x, y) = d(y, x),$$

$$d(x, y) = 0 \iff x = y$$

84 Trees

and

$$d(x, z) \leq d(x, y) + d(y, z).$$

We also say that d *is a metric on S*.

The last clause in the definition is called the *triangle inequality*. The most important example of a metric space (S, d) has

$$S = \mathbb{R}$$

and

$$d = |x - y|.$$

In this context, $|x - y|$ means the absolute value of the difference between x and y. This is the usual distance function for $\mathbb{R}$. Here are some well-known facts:

$$|x - y| \geq 0,$$

$$|x - y| = |y - x|,$$

$$|x - y| = 0 \iff x = y$$

and

$$|x - z| \leq |x - y| + |y - z|.$$

The following lemma is immediate from these facts.

Lemma 5.4 *The usual distance function for $\mathbb{R}$ is a metric on $\mathbb{R}$.*

Notice that each open interval (a, b) of $\mathbb{R}$ has the form

$$\{x \in \mathbb{R} \mid |x - c| < r\}$$

for some $c \in \mathbb{R}$ (the center) and positive $r \in \mathbb{R}$ (the radius). Just take $c = (b + a)/2$ and $r = (b - a)/2$ to see this. In this sense, the topology of $\mathbb{R}$ comes from the metric on $\mathbb{R}$. One says that the topology and the metric are *compatible* when they are related in this manner. Not every topological space has a compatible metric but many interesting ones do.

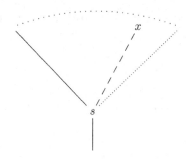

Figure 5.1 A basic open neighborhood N_s and $x \in N_s$

5.2 The Baire space

In this section, we endow ${}^{\omega}\omega$ with a topology and a metric, which turn out to be compatible. Throughout this and subsequent sections, it is very important to keep in mind the distinction between ${}^{<\omega}\omega$ (the set of finite sequences of natural numbers) and ${}^{\omega}\omega$ (the set of infinite sequences of natural numbers).

Definition 5.5 If $n < \omega$ and $s \in {}^{n}\omega$, then

$$N_s = \{x \in {}^{\omega}\omega \mid x \restriction n = s\}.$$

These are the *basic open subsets of* ${}^{\omega}\omega$.

Figure 5.1 is an attempt to illustrate the basic open set N_s, which consists of all the infinite branches x that pass through s. The following easy observation is often useful.

Lemma 5.6 *If $s, t \in {}^{<\omega}\omega$, then*

- *if $s \subseteq t$, then $N_t \subseteq N_s$,*
- *if $t \subseteq s$, then $N_s \subseteq N_t$, and*
- *otherwise, $N_s \cap N_t = \emptyset$.*

Definition 5.7 U *is an* open *subset of* ${}^{\omega}\omega$ *iff there is a family* $\mathcal{F}$ of basic open subsets of ${}^{\omega}\omega$ such that

$$U = \bigcup \mathcal{F}$$

Note that, because $^{<\omega}\omega$ is countable, U is an open subset of $^\omega\omega$ iff there is a sequence $\langle s_n \mid n < \omega \rangle$ from $^{<\omega}\omega$ such that

$$U = \bigcup_{n<\omega} N_{s_n}.$$

Lemma 5.8 $\{U \mid U$ *is an open subset of* $^\omega\omega\}$ *is a topology on* $^\omega\omega$.

This is the *Baire topological space*.

Proof   Everything is obvious except the fact that the intersection of finitely many open sets is open. It is enough to show that the intersection of two open sets is open. The general statement follows by induction because we can add parentheses as follows:

$$U_0 \cap U_1 \cap \cdots \cap U_n = U_0 \cap (U_1 \cap \cdots \cap U_n).$$

Say

$$A = \bigcup \mathcal{F}$$

and

$$B = \bigcup \mathcal{G},$$

where $\mathcal{F}$ and $\mathcal{G}$ are families of basic open sets. We must show that $A \cap B$ is an open set. Let $\mathcal{H}$ be the collection of basic open sets N_t for which there are $r, s \in {}^{<\omega}\omega$ such that

- $N_r \in \mathcal{F}$ and $N_s \in \mathcal{G}$,
- $r \subseteq s$ or $s \subseteq r$, and
- $t = r \cup s$.

The last two clauses say that either

$$r = s \restriction \mathrm{dom}(r)$$

or

$$s = r \restriction \mathrm{dom}(s)$$

(we say r and s are *comparable* in this case), and t is the longer of the two (which is the union because they are comparable). We will be done when we show that

$$A \cap B = \bigcup \mathcal{H}.$$

First suppose that $x \in A \cap B$. Then there are $r, s \in {}^{<\omega}\omega$ such that

$$x \in N_r \in \mathcal{F}$$

and

$$x \in N_s \in \mathcal{G}.$$

Then the finite sequences r and s are comparable because they are both restrictions of the same infinite sequence x. That is,

$$r = x \upharpoonright \mathrm{dom}(r)$$

and

$$s = x \upharpoonright \mathrm{dom}(s).$$

Let $t = r \cup s$ be the longer of the two. Then

$$x \in N_t \in \mathcal{H},$$

so

$$x \in \bigcup \mathcal{H}.$$

This shows that

$$A \cap B \subseteq \bigcup \mathcal{H}.$$

We leave the easier reverse inclusion to the reader. $\qquad\square$

Definition 5.9 C is a *closed* subset of ${}^{\omega}\omega$ iff ${}^{\omega}\omega - C$ is an open subset of ${}^{\omega}\omega$.

It is time for some examples. Consider an arbitrary $x \in {}^{\omega}\omega$. The singleton $\{x\}$ is closed since its complement is open:

$$ {}^{\omega}\omega - \{x\} = \bigcup \{ N_s \mid s \in {}^{<\omega}\omega \text{ but } s \not\subseteq x \}.$$

However, $\{x\}$ is not open since, for every $s \in {}^{<\omega}\omega$,

$$N_s \not\subseteq \{x\}$$

because N_s has infinitely many elements while $\{x\}$ has just one. Thus $\{x\}$ is closed but not open. It follows easily that ${}^{\omega}\omega - \{x\}$ is open but not closed.

Most sets are neither open nor closed. One way to see this is to observe that

$$\left| \{ N_s \mid s \in {}^{<\omega}\omega \} \right| = \left| {}^{<\omega}\omega \right| = \aleph_0,$$

and

$$|\{C \mid C \text{ is a closed subset of } {}^\omega\omega\}| = |\{U \mid U \text{ is an open subset of } {}^\omega\omega\}|$$
$$= \left|{}^\omega\left({}^{<\omega}\omega\right)\right|$$
$$= \aleph_0^{\aleph_0}$$
$$= 2^{\aleph_0},$$

which is strictly smaller than

$$|\mathcal{P}({}^\omega\omega)| = 2^{2^{\aleph_0}}.$$

Since there are strictly more subsets of the Baire space than there are open or closed subsets, there must be subsets which are neither open nor closed.

It is also easy to come up with specific examples of sets which are neither open nor closed. Given $n < \omega$, $s \in {}^n\omega$ and $x \in {}^\omega\omega$, let

$$s^\frown x = s \cup \{(n+k, x(k)) \mid k < \omega\}.$$

A less precise but somehow clearer way to write this is

$$s^\frown x = \langle s(0), \ldots, s(n-1), x(0), x(1), x(2), \ldots \rangle$$

where if $n = 0$, then $s^\frown x = x$. Suppose that U is open but not closed and C is closed but not open. (We already gave examples of such sets.) Let

$$A = \{\langle 0 \rangle^\frown x \mid x \in U\} \cup \{\langle 1 \rangle^\frown y \mid y \in C\}.$$

We will prove that A is not open and leave the verification that A is not closed to the reader. As C is not open, there exists $x \in C$ such that, for every $n < \omega$,

$$N_{x \restriction n} \not\subseteq C.$$

Let

$$y = \langle 1 \rangle^\frown x.$$

Then, for every $n < \omega$,

$$N_{y \restriction n} \not\subseteq A.$$

This implies that A is not open.

Definition 5.10 *Clopen* means both closed and open.

If $(S, \mathcal{T})$ is a topological space, then $S \in \mathcal{T}$ by definition and $\emptyset \in \mathcal{T}$ because $\emptyset = \bigcup \emptyset$ is the union of the empty family of open sets. Thus both $\emptyset$ and S are always clopen. In the standard topology on $\mathbb{R}$, the only clopen sets are $\emptyset$ and $\mathbb{R}$. In the Baire space, there are clopen sets other than $\emptyset$ and $^\omega\omega$. For example, if $n < \omega$ and $s \in {}^n\omega$, then N_s is clopen since N_s is obviously open and

$$^\omega\omega - N_s = \bigcup\{N_t \mid t \in {}^n\omega \text{ but } t \neq s\}$$

is also open. We get more examples by noting that a union of finitely many clopen sets is also clopen. For example,

$$N_{\langle 0 \rangle} \cup N_{\langle 1 \rangle} = \{x \in {}^\omega\omega \mid x(0) = 0 \text{ or } x(0) = 1\}$$

is clopen.

Definition 5.11 For $x, y \in {}^\omega\omega$, the distance between x and y is

$$d(x, y) = \begin{cases} 1/2^n & \text{if } n \text{ is least such that } x(n) \neq y(n) \\ 0 & \text{if } x = y. \end{cases}$$

For example,

$$d\left(\langle 0, 7, 4, 3, \ldots \rangle, \langle 0, 7, 9, 9, \ldots \rangle\right) = 1/2^2 = 1/4.$$

The proofs of the following lemmas are left to the reader.

Lemma 5.12 *d is a metric on $^\omega\omega$.*

Lemma 5.13 *For every $A \subseteq {}^\omega\omega$, the following are equivalent.*

1. *There exists $s \in {}^{<\omega}\omega$ such that $A = N_s$.*
2. *There exist $c \in {}^\omega\omega$ and a real number $r > 0$ such that*

$$A = \{x \in {}^\omega\omega \mid d(x, c) < r\}.$$

The previous lemma says that the topology on $^\omega\omega$ is compatible with the metric d. It is worth observing that d is not the only metric with this property. For example, if we define e so that

$$e(x, y) = \begin{cases} 1/(n + 1) & \text{if } n \text{ is least such that } x(n) \neq y(n) \\ 0 & \text{if } x = y, \end{cases}$$

then e is also a metric compatible with the Baire topology on $^\omega\omega$.

It is also worth observing that, although we wrote c for *center* and r for *radius* in Lemma 5.13, the Baire metric space is different from the real line in that the center and radius of a basic open set are not unique. For example,

$$\{x \in {}^\omega\omega \mid d(x, c) < r\} = N_{\langle\ \rangle} = {}^\omega\omega$$

for every $c \in {}^\omega\omega$ and real number $r > 1$. We are using the notation $\langle\ \rangle$ to denote the empty sequence. Technically, $\langle\ \rangle = \emptyset = 0$, so we have three names for the same thing. Another example is

$$\{x \in {}^\omega\omega \mid d(x, c) < r\} = N_{\langle 0 \rangle}$$

for every $c \in {}^\omega\omega$ with $c(0) = 0$ and $1/2 < r \leq 1$.

Next we explain what this has to do with trees, the title of this chapter. The following definition of *tree* is not the most general but it suffices for all but the last section of this chapter.

Definition 5.14   Let Ω be a set. Then T is a *tree on* Ω iff

$$T \subseteq {}^{<\omega}\Omega$$

and, for all $m, n \in \omega$ and $s \in {}^n\omega$, if $s \in T$ and $m < n$, then $s \restriction m \in T$.

We will focus on the cases $\Omega = \omega$ and $\Omega = 2$. As an example,

$$T = \{\langle\ \rangle, \langle 2 \rangle, \langle 7 \rangle, \langle 2, 8 \rangle, \langle 2, 9 \rangle, \langle 7, 1 \rangle, \langle 7, 5 \rangle, \langle 7, 7 \rangle,$$
$$\langle 2, 8, 1 \rangle, \langle 2, 8, 1, 1 \rangle, \langle 2, 8, 1, 5 \rangle\}$$

is a tree on ω. As part of checking that T is a tree, note that $\langle 2, 8, 1, 5 \rangle \in T$ and so are all of its restrictions: $\langle 2, 8, 1 \rangle$, $\langle 2, 8 \rangle$, $\langle 2 \rangle$ and $\langle\ \rangle$.

Definition 5.15   If T is a tree on ω, then the *set of infinite branches* of T is

$$[T] = \{x \in {}^\omega\omega \mid x \restriction n \in T \text{ for every } n < \omega\}.$$

Example $[{}^{<\omega}\omega] = {}^\omega\omega$.

Example $[{}^{<\omega}2] = {}^\omega 2$.

Example   If $s \in {}^{<\omega}\omega$, then $[\{r \in {}^{<\omega}\omega \mid r \subseteq s \text{ or } s \subseteq r\}] = N_s$.

Example   If $x \in {}^\omega\omega$, then $[\{x \restriction n \mid n < \omega\}] = \{x\}$.

It is easy to see that these examples of sets of the form $[T]$ are closed subsets of the Baire space. This is no accident as the following result explains.

Lemma 5.16 *Let $C \subseteq {}^\omega\omega$. Then C is a closed subset of ${}^\omega\omega$ iff there is a tree T on ω such that $C = [T]$.*

Proof First we prove the reverse direction. Assume T is a tree on ω and $C = [T]$. Let $U = {}^\omega\omega - C$. To see that C is closed we show that U is open. For this, simply observe that

$$U = {}^\omega\omega - [T]$$
$$= \{x \in {}^\omega\omega \mid \text{there exists } n < \omega \text{ such that } x \restriction n \notin T\}$$
$$= \bigcup\{N_s \mid s \in {}^{<\omega}\omega - T\}$$

is a union of basic open sets.

For the forward direction of Lemma 5.16, consider an arbitrary closed subset C of ${}^\omega\omega$. Put

$$T = \{x \restriction n \mid n < \omega \text{ and } x \in C\}.$$

Clearly $C \subseteq [T]$. We finish by showing that $[T] \subseteq C$. For contradiction, suppose

$$y \in [T] - C.$$

Let $U = {}^\omega\omega - C$. Then U is open and $y \in U$. So there exists $n < \omega$ such that

$$N_{y \restriction n} \subseteq U.$$

In other words,

$$N_{y \restriction n} \cap C = \emptyset.$$

But, since $y \in [T]$, there exists $x \in C$ such that $x \restriction n = y \restriction n$. Thus

$$x \in N_{y \restriction n} \cap C.$$

This contradiction completes the proof. $\qquad\square$

Exercises

Exercise 5.1 Let

$$I = \{x \in {}^\omega\omega \mid x \text{ is an injection from } \omega \text{ to } \omega\}$$

and

$$S = \{x \in {}^\omega\omega \mid x \text{ is a surjection from } \omega \text{ to } \omega\}.$$

Answer the following questions and prove your answer is correct.

1. Is I open?
2. Is I closed?
3. Is S open?
4. Is S closed?

Exercise 5.2   Let $A \subseteq {}^{\omega}\omega$. Put

$$T = \{x \restriction n \mid n < \omega \text{ and } x \in A\}.$$

Prove that $[T]$ is the *closure of A* in the Baire space. By this we mean that $[T]$ is closed and, for every closed set C, if $A \subseteq C$, then $[T] \subseteq C$.

Exercise 5.3   A topological space $(S, \mathcal{T})$ is said to be a *Lindelöf space* iff for every $\mathcal{F} \subseteq \mathcal{T}$, if

$$S = \bigcup \mathcal{F},$$

then there is a countable $\mathcal{G} \subseteq \mathcal{F}$ such that

$$S = \bigcup \mathcal{G}.$$

Prove that the Baire space is a Lindelöf space.

Exercise 5.4   A topological space $(S, \mathcal{T})$ is said to be *compact* iff for every $\mathcal{F} \subseteq \mathcal{T}$, if

$$S = \bigcup \mathcal{F},$$

then there is a finite $\mathcal{G} \subseteq \mathcal{F}$ such that

$$S = \bigcup \mathcal{G}.$$

1. Prove that the Baire space is not compact.
2. The *Cantor space* is the topological space on ${}^{\omega}2$ whose open sets are exactly those of the form ${}^{\omega}2 \cap U$ where U is an open subset of the Baire space. You could say that the Cantor space topology is *inherited* from the Baire space.

   Prove that the Cantor space is compact.

   Hint: Let $\mathcal{F}$ be a family of open subsets of ${}^{\omega}2$. Assume there is no finite $\mathcal{G} \subseteq \mathcal{F}$ such that

   $$\bigcup \mathcal{G} = {}^{\omega}2.$$

Prove that there is an $x \in {}^{\omega}2$ such that

$$x \notin \bigcup \mathcal{F}.$$

Use recursion to define $x(n)$ in terms of $x \upharpoonright n$. Along the recursion, maintain that there is no finite $\mathcal{G} \subseteq \mathcal{F}$ such that

$$\bigcup \mathcal{G} \supseteq N_{x \upharpoonright n} \cap {}^{\omega}2.$$

Exercise 5.5   If $(S, \mathcal{T})$ is a topological space and $D \subseteq S$, then D is said to be *dense* iff for every non-empty $U \in \mathcal{T}$,

$$D \cap U \neq \emptyset.$$

A topological $(S, \mathcal{T})$ space is said to be *separable* iff it has a countable dense subset. Show that the Baire space is separable.

Exercise 5.6   Let D be the set of $x \in {}^{\omega}\omega$ such that, for every $m < \omega$, there exists $n < \omega$ such that $m < n$ and $x(n) = 0$.

1. Prove that D is dense.
2. Prove that D is not open.
3. Prove that D is not closed.
4. Find a sequence $\langle U_n \mid n < \omega \rangle$ of subsets of ${}^{\omega}\omega$ such that

$$D = \bigcap_{n < \omega} U_n$$

   and, for every $n < \omega$, U_n is open and dense.

Exercise 5.7   We need three definitions before stating the exercise. Consider an arbitrary metric space (S, d).

- Let $\langle x_i \mid i < \omega \rangle$ be a sequence of elements of S and $y \in S$.

  - We say that $\langle x_i \mid i < \omega \rangle$ *converges to* y and write

    $$\lim_{i \to \omega} x_i = y$$

    iff for every $r \in \mathbb{R}$, if $r > 0$, then there exists $i < \omega$ such that, for every $j < \omega$, if $j > i$, then $d(x_j, y) < r$.
  - We call $\langle x_i \mid i < \omega \rangle$ a *Cauchy sequence* iff for every $r \in \mathbb{R}$, if $r > 0$, then there exists $i < \omega$ such that, for all $j, k < \omega$, if $j, k > i$, then $d(x_j, x_k) < r$.

- We say that (S, d) is *complete* iff for every Cauchy sequence $\langle x_i \mid i < \omega \rangle$ from S, there is $y \in S$ such that

$$\lim_{i \to \omega} x_i = y.$$

The following exercises are about the Baire space with the metric defined by

$$d(x, y) = 1/2^n \iff (x \upharpoonright n = y \upharpoonright n \text{ but } x(n) \neq y(n))$$

and

$$d(x, y) = 0 \iff x = y$$

but we remark that parts 1 and 2 hold in every metric space.

1. Let $C \subseteq {}^\omega \omega$. Prove that C is closed iff for every sequence

$$\langle x_i \mid i < \omega \rangle$$

from C and every $y \in S$, if

$$\lim_{i \to \omega} x_i = y,$$

then $y \in C$. This says that a set is closed iff it has all its limit points.
2. Prove that if $\lim_{i \to \omega} x_i = y$, then $\langle x_i \mid i < \omega \rangle$ is a Cauchy sequence.
3. Prove that the Baire space is complete.

Exercise 5.8 (Baire category theorem)   Let $\langle D_n \mid n < \omega \rangle$ be a sequence of subsets of ${}^\omega \omega$. Assume that, for every $n < \omega$, D_n is both open and dense in the Baire space. Let

$$E = \bigcap_{n < \omega} D_n.$$

Prove that E is dense in the Baire space.

Exercise 5.9   A tree T on ω is called *perfect* iff for every $r \in T$, there are $s, t \in T$ such that $r \subseteq s$, $r \subseteq t$, $s \not\subseteq t$ and $t \not\subseteq s$. Prove that if T is a non-empty perfect tree on ω, then T has $2^{\aleph_0}$ many branches, that is,

$$\|[T]\| = 2^{\aleph_0}.$$

Exercise 5.10 Let $\langle D_n \mid n < \omega \rangle$ be a sequence of subsets of ${}^\omega \omega$. Assume that, for every $n < \omega$, D_n is both open and dense in the Baire space. Let

$$E = \bigcap_{n<\omega} D_n.$$

Prove that

$$|E| = 2^{\aleph_0}.$$

Hint: By Exercise 5.9, it is enough to show that there is a perfect tree T such that $[T] \subseteq E$. Construct T using ideas similar to the solution to Exercise 5.8.

Exercise 5.11 (Cantor perfect set theorem) Let C be a closed subset of the Baire space and T be a tree on ω such that $C = [T]$. The *Cantor–Bendixon derivative* of T is defined to be

$$T' = \{s \in T \mid N_s \cap [T] \text{ has at least two elements}\}.$$

By recursion, define

$$T^0 = T,$$

$$T^{\alpha+1} = (T^\alpha)'$$

whenever α is an ordinal, and

$$T^\beta = \bigcap_{\alpha<\beta} T^\alpha.$$

whenever β is a limit ordinal.

1. By induction on all ordinals β, prove that T^β is a tree on ω and, for every $\alpha < \beta$,

$$T^\beta \subseteq T^\alpha.$$

2. Prove that there exists $\delta < \omega_1$ such that $T^{\delta+1} = T^\delta$.
3. Let δ be least such that

$$T^{\delta+1} = T^\delta.$$

(This is the *Cantor–Bendixon rank* of T.)

(a) Prove that T^δ is a perfect tree and $[T^\delta] \subseteq C$.
(b) Prove that if $T^\delta = \emptyset$, then $|C| \leq \aleph_0$.

Notice that the combination of Exercises 5.9 and 5.11 shows that closed subsets of the Baire space are either countable or have cardinality $2^{\aleph_0}$.

Exercise 5.12 Prove by induction that, for every $\delta < \omega_1$, there is a tree T_δ on ω whose Cantor–Bendixon rank is δ and $(T_\delta)^\delta = \emptyset$. *Hint*: Obviously, $T_0 = \emptyset$ and $T_1 = {}^{<\omega}\{0\}$ work. Next define T_2 and T_3. Once you see a pattern for natural numbers, try T_ω. Then you will be on your way to constructing T_δ by recursion.

Exercise 5.13 Prove that there exists a set $A \subseteq {}^\omega\omega$ such that, for every non-empty perfect tree T, neither $[T] \subseteq A$ nor $[T] \subseteq {}^\omega\omega - A$. *Hint*: Let

$$\langle T_\alpha \mid \alpha < 2^{\aleph_0} \rangle$$

enumerate the non-empty perfect trees. Recursively define

$$\langle x_\alpha \mid \alpha < 2^{\aleph_0} \rangle$$

and

$$\langle y_\alpha \mid \alpha < 2^{\aleph_0} \rangle$$

such that, for every $\beta < 2^{\aleph_0}$,

$$y_\beta \in [T_\beta] - \{x_\alpha \mid \alpha < \beta\}$$

and

$$x_\beta \in [T_\beta] - \{y_\alpha \mid \alpha \leq \beta\}.$$

Then let $A = \{x_\alpha \mid \alpha < 2^{\aleph_0}\}$.

   Remark: Exercise 5.13 is another example of a diagonal argument. Intuitively, we diagonalize over all non-empty perfect trees to make sure none of them work.

5.3 Illfounded and wellfounded trees

The first result of this section is that if T is a tree on ω with infinite height and finite levels, then T has an infinite branch.

Theorem 5.17 (D. König) *Let T be a tree on ω. Assume that, for every $n < \omega$,*

$$T \cap {}^n\omega \neq \emptyset$$

and

$$|T \cap {}^n\omega| < \aleph_0.$$

Then

$$[T] \neq \emptyset.$$

Corollary 5.18 *Let T be a tree on 2. Assume that, for every $n < \omega$,*

$$T \cap {}^n 2 \neq \emptyset.$$

Then

$$[T] \neq \emptyset.$$

Corollary 5.18 is an immediate consequence of Theorem 5.17, which we will prove after some discussion and an example. We already used the words *level* and *height* informally. Now let us officially define them.

Definition 5.19 Let T be a tree on ω. Then, for every $n < \omega$,

$$\mathrm{level}_n(T) = T \cap {}^n\omega$$

and

$$\mathrm{height}(T) = \{n < \omega \mid \mathrm{level}_n(T) \neq \emptyset\}.$$

Notice that if T is a tree on ω, then $\mathrm{height}(T)$ is an ordinal and

$$\mathrm{height}(T) \leq \omega.$$

This is because trees are closed downward: if $s \in \mathrm{level}_n(T)$ and $m < n$, then $s \restriction m \in \mathrm{level}_m(T)$. As $\mathrm{height}(T)$ is a transitive set of natural numbers, it is itself an ordinal $\leq \omega$.

In Theorem 5.17, we cannot drop the hypothesis that all levels of T are finite. Consider the tree depicted in Figure 5.2. It consists of all restrictions $s_n \restriction m$ of sequences

$$s_n = \langle n, 0, \ldots, 0 \rangle$$

where, in the displayed sequence, $n < \omega$ and there are n zeros. Then T has infinite height but no infinite branch.

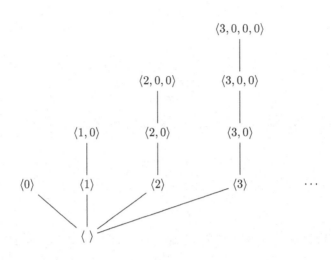

Figure 5.2 An infinite tree with no infinite branches

Proof of Theorem 5.17   We will need the following notation. Given $r \in {}^{<\omega}\omega$, let

$$T_r = \{s \in T \mid r \subseteq s \text{ or } s \subseteq r\}.$$

Notice that $T_r \subseteq T$, T_r is a tree on ω and T_r has finite levels.

Define a function $x : \omega \to \omega$ by recursion as follows. Assume that $x \upharpoonright n$ has been defined so that $x \upharpoonright n \in T$ and $T_{x \upharpoonright n}$ has infinite height. Since $T_{x \upharpoonright n}$ has finite levels, we can write

$$T_{x \upharpoonright n} = \bigcup_{i<j} T_{s_i}$$

where $j < \omega$ and, for every $i < j$,

$$s_i \in {}^{n+1}\omega \cap T_{x \upharpoonright n}.$$

From the equation above and the fact that $T_{x \upharpoonright n}$ has infinite height, it follows that there is at least one $i < j$ such that T_{s_i} has infinite height. Define $x \upharpoonright (n+1) = s_i$ for the least such i. In other words, put $x(n) = s_i(n)$ for this i. This completes the definition of x. Clearly, $x \in [T]$, which proves the theorem. $\square$

Recall that the Cantor space is compact by Exercise 5.4. It is possible to derive Corollary 5.18 directly from the fact that the Cantor space is compact. This alternative proof uses Lemma 5.16, which, in the case of the Cantor space, tells us that C is closed subset of $^\omega 2$ iff there is a tree T on 2 such that $C = [T]$. Consider this a hint for Exercise 5.15.

Theorem 5.17 gives conditions that imply a tree has infinite branches. Now we want to understand when a tree does not have infinite branches. It may help to picture trees growing downward instead of upward for this discussion. Not having infinite branches makes a tree wellfounded according to the following definition.

Definition 5.20 If T is a tree on ω, then T is *wellfounded* iff $[T] = \emptyset$. Otherwise, T is *illfounded*. We call s a *terminal node of* T iff $s \in T$ and there is no $t \in T$ with $t \supsetneq s$.

This terminology makes sense if you think of the tree as growing downward because if $x \in [T]$, then

$$\cdots \supsetneq x \upharpoonright 2 \supsetneq x \upharpoonright 1 \supsetneq x \upharpoonright 0.$$

Theorem 5.17 says that an infinite tree with finite levels is ill-founded. We will characterize wellfounded trees in terms of rank functions.

Definition 5.21 Let T be a tree on ω. A *rank function* for T is function f with domain T such that, for all $s, t \in T$,

- $f(s)$ is an ordinal and
- if $s \subsetneq t$, then $f(s) > f(t)$.

Lemma 5.22 *If T is a tree on ω and T has a rank function, then T is wellfounded.*

Proof For contradiction, suppose that $[T] \neq \emptyset$. Let $x \in [T]$. Then

$$\cdots < f(x \upharpoonright 2) < f(x \upharpoonright 1) < f(x \upharpoonright 0)$$

is an infinite descending sequence of ordinals. $\quad\square$

The converse of Lemma 5.22 is also true. In fact, if T is a well-founded tree on ω, then there is a natural way to define a rank function for T, which is what the following theorem explains.

Theorem 5.23   *Let T be a wellfounded tree on ω. Then there is a unique function*

$$\rho_T : T \longrightarrow \omega_1$$

such that, for every $s \in T$,

- *if s is a terminal node of T, then $\rho_T(s) = 0$, and*
- *if s is not a terminal node of T, then*

$$\rho_T(s) = \sup\left(\{\rho_T(t) + 1 \mid t \supsetneq s\}\right).$$

In particular, ρ_T is a rank function for T.

We call ρ_T the *rank function associated to T*. The two clauses determining ρ_T in Theorem 5.23 look like a recursive definition but it is not immediately clear which wellordering underlies the recursion. This is sorted out in Exercises 5.18 and 5.19.

Corollary 5.24   *Let T be a tree on ω. Then T is wellfounded iff T has a rank function.*

Definition 5.25   *If T is a non-empty wellfounded tree on ω, then we let the rank of T be*

$$\mathrm{rank}(T) = \rho_T(\langle\,\rangle).$$

Do not confuse height with rank. Every tree on ω has height at most ω. But only wellfounded trees have ranks, and these ranks are sometimes strictly greater than ω.

Example   For every $n < \omega$, the trees

$$\{s \mid \mathrm{dom}(s) \le n \text{ and } s(m) = 1 \text{ for every } m < \mathrm{dom}(s)\}$$

and

$$\{s \mid \mathrm{dom}(s) \le n \text{ and } s(m) < \omega \text{ for every } m < \mathrm{dom}(s)\}$$

both have rank n.

Example   Figure 5.2 shows an example of a wellfounded tree of rank ω. Figure 5.3 shows the same tree with the nodes labeled according to their rank values.

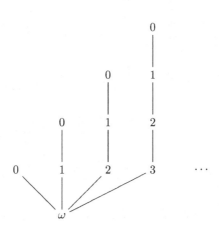

Figure 5.3 A wellfounded tree of rank ω (labels are ranks)

Example If we let T be the tree in Figure 5.2 and define

$$U = \{\langle 0 \rangle ^\frown s \mid s \in T\},$$

then

$$\operatorname{rank}(U) = \omega + 1.$$

See Figure 5.4 for a picture of U with its nodes labeled according to their rank values. We should explain the notation we are using in the definition of U. For $r \in {}^m\omega$ and $s \in {}^n\omega$, let

$$r^\frown s = \langle r(0), \ldots, r(m-1), s(0), \ldots, s(n-1) \rangle \in {}^{m+n}\omega.$$

Put another way,

$$(r^\frown s)(i) = \begin{cases} r(i) & \text{if } i < m \\ s(i-m) & \text{if } m \le i < m+n. \end{cases}$$

The examples above beg the question: which ordinals are the ranks of trees on ω? The answer is exactly the countable ordinals by Theorem 5.23 and Exercise 5.20.

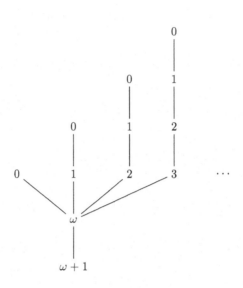

Figure 5.4 A wellfounded tree of rank $\omega + 1$ (labels are ranks)

Exercises

Exercise 5.14 Let T be a tree on ω. Assume that $[T] \neq \emptyset$. Prove that there exists a unique $x \in [T]$ such that, for all $y \in [T]$ and $n < \omega$, if $y \upharpoonright n = x \upharpoonright n$, then $x(n) \leq y(n)$. We call x the *left-most branch of* T.

Exercise 5.15 You should notice that the proof of Theorem 5.17 is similar to the solution to Exercise 5.4(2). This exercise explains why.

1. Use Lemma 5.16 and Corollary 5.18 to derive the fact that the Cantor space is compact.
2. Use Lemma 5.16 and the fact that the Cantor space is compact to derive Corollary 5.18.

Exercise 5.16 Recall that *clopen* means closed and open.

1. Prove that, for every $C \subseteq {}^\omega 2$, if C is a clopen subset of the Cantor space, then C is a union of finitely many basic open subsets of the Cantor space. In other words, there is a finite set

$$\{s_0, \ldots, s_{n-1}\} \subset {}^{<\omega}\omega$$

such that

$$C = \bigcup_{i<n} N_{s_i} \cap {}^\omega 2.$$

2. Find an example of a set C such that C is a clopen subset of the Baire space but neither C nor ${}^\omega\omega - C$ is a finite union of basic open sets. Explain why your example has this property.
3. Prove that the only two clopen subsets of $\mathbb{R}$ are $\emptyset$ and $\mathbb{R}$.

Exercise 5.17 Let T be the wellfounded tree consisting of descending sequences of natural numbers. In other words,

$$T = \{s \in {}^{<\omega}\omega \mid \text{for all } m, n \in \mathrm{dom}(s), \text{ if } m < n, \text{ then } s(m) > s(n)\}.$$

Calculate $\mathrm{rank}(T)$.

Exercise 5.18 (Kleene–Brouwer ordering) Define a relation $<_{\mathrm{KB}}$ on ${}^{<\omega}\omega$ by declaring that, for all $s, t \in {}^{<\omega}\omega$,

$$t <_{\mathrm{KB}} s$$

iff either $t \supsetneq s$ or there exists $n < \omega$ such that $t \upharpoonright n = s \upharpoonright n$ but $t(n) < s(n)$.

1. Prove that $<_{\mathrm{KB}}$ is a strict linear ordering of ${}^{<\omega}\omega$.
2. Let T be a tree on ω.

    (a) Prove that if the restriction of $<_{\mathrm{KB}}$ to T is a wellordering, then T is wellfounded.
    (b) Prove that if T is wellfounded, then the restriction of $<_{\mathrm{KB}}$ to T is a wellordering.

Exercise 5.19 Prove Theorem 5.23 in the following two steps.

1. Explain why the properties of ρ_T listed in the statement of Theorem 5.23 form a legitimate definition by recursion on the restriction of $<_{\mathrm{KB}}$ to T.
2. Explain why the range of ρ_T is a countable ordinal.

Exercise 5.20   Prove by induction on $\alpha < \omega_1$ that there exists a wellfounded tree T on ω with $\mathrm{rank}(T) = \alpha$.

Exercise 5.21   Let T be a non-empty wellfounded tree on 2. Prove that $\mathrm{rank}(T) < \omega$.

Exercise 5.22   Let $\mathbb{B}$ be the Boolean algebra of clopen subsets of the Cantor space. That is,

$$B = \{X \mid X \text{ is a clopen subset of } {}^{\omega}2\},$$

$$X \vee_{\mathbb{B}} Y = X \cup Y,$$

$$X \wedge_{\mathbb{B}} Y = X \cap Y,$$

$$\neg_{\mathbb{B}} X = {}^{\omega}2 - X,$$

$$\bot_{\mathbb{B}} = \emptyset$$

and

$$\top_{\mathbb{B}} = {}^{\omega}2.$$

Prove $\mathbb{B}$ is a countable atomless Boolean algebra.

Exercise 5.23   Let $\mathbb{B}$ be the Boolean algebra of clopen subsets of the Baire space. That is,

$$B = \{X \mid X \text{ is a clopen subset of } {}^{\omega}\omega\},$$

$$X \vee_{\mathbb{B}} Y = X \cup Y,$$

$$X \wedge_{\mathbb{B}} Y = X \cap Y,$$

$$\neg_{\mathbb{B}} X = {}^{\omega}\omega - X,$$

$$\bot_{\mathbb{B}} = \emptyset$$

and

$$\top_{\mathbb{B}} = {}^{\omega}\omega.$$

Prove $\mathbb{B}$ is an atomless Boolean algebra of cardinality $2^{\aleph_0}$.

5.4 Infinite games

Let $A \subseteq {}^{\omega}\omega$. We describe a game, which is called G_A. The game has two players, I and II, who take turns playing natural numbers x_0, x_1, etc. A run of the game G_A looks as follows.

I x_0 x_2 x_4

                                $\ldots$

II x_1 x_3 x_5

If $x = \langle x_n \mid n < \omega \rangle$ is a run of G_A, then player I wins the run iff $x \in A$. Otherwise, player II wins the run.

This is a very general sort of game. Notice that finite games also fit this scheme because we may ignore moves after a winner has been declared. (There are really two kinds of finite length games. Either the length is fixed in advance or else the length depends on exactly how the players move. Both kinds of finite games can be modeled with our infinite games.) One difference between our games and some familiar games like chess is that we do not allow a run of the game to end in a draw. This is because either $x \in A$ or $x \notin A$. An arbitrary way to get around this objection is to declare that draws go to player II. See Exercise 5.24 for more about chess.

Many properties in mathematics can be expressed in terms of games, so general theorems about games can be quite useful. Along these lines, a series of exercises at the end of this section illustrate one of many ways in which games and mathematical analysis are related.

Naturally, we are more interested in winning strategies than we are in the player who wins a particular run of a particular game. So let us continue making definitions associated to the game G_A.

A *strategy* is a function $\sigma : {}^{<\omega}\omega \to \omega$. If σ is a strategy and $b \in {}^{\omega}\omega$, then $\sigma * b$ is the run that results if I uses σ and II plays b. Formally, $\sigma * b$ is defined by recursion according to the equations

$$(\sigma * b)_{2n+1} = b_n$$

and

$$(\sigma * b)_{2n} = \sigma((\sigma * b) \restriction 2n).$$

See Figure 5.5 for another way of depicting the run $\sigma * b$. We call σ a *winning strategy for player I* iff for every $b \in {}^{\omega}\omega$,

$$\sigma * b \in A.$$

I $\sigma(\langle\ \rangle)$ $\sigma(\langle\sigma(\langle\ \rangle),b_0\rangle)$ $\sigma(\langle\sigma(\langle\ \rangle),b_0,\sigma(\langle\sigma(\langle\ \rangle),b_0\rangle),b_1\rangle)$

II b_0 b_1 b_2 $\cdots$

Figure 5.5 *The run* $\sigma * b$

I a_0 a_1 a_2 a_3 $\cdots$

II $\sigma(\langle a_0 \rangle)$ $\sigma(\langle a_0, \sigma(\langle a_0 \rangle), a_1 \rangle)$ $\sigma(\langle a_0, \sigma(\langle a_0 \rangle), a_1, \sigma(\langle a_0, \sigma(\langle a_0 \rangle), a_1 \rangle), a_2 \rangle)$

Figure 5.6 *The run* $a * \sigma$

Similarly, if σ is a strategy and $a \in {}^\omega\omega$, then $a * \sigma$ is the run that results if II uses σ and I plays a. Formally, $a * \sigma$ is defined by recursion according to the equations

$$(a * \sigma)_{2n} = a_n$$

and

$$(a * \sigma)_{2n+1} = \sigma((a * \sigma) \upharpoonright (2n + 1)).$$

Figure 5.6 depicts the run $a * \sigma$ another way. We call σ a *winning strategy for player II* iff for every $a \in {}^\omega\omega$,

$$a * \sigma \notin A.$$

We say that A is *determined* iff either player I has a winning strategy or player II has a winning strategy. Otherwise, A is *undetermined*. Obviously, it is not possible for both players to have winning strategies because otherwise we could play the two strategies against each other to get a contradiction.

Example Let A be the set of surjections from ω to ω. Let σ be the strategy such that

$$\sigma(\langle x_i \mid i < 2n \rangle) = n$$

and

$$\sigma(\langle x_i \mid i < 2n + 1 \rangle) = 0.$$

Then, for every $b \in {}^\omega\omega$,

$$\sigma * b \in A.$$

Thus σ is a winning strategy for player I in G_A. So A is determined. Observe that σ is not the only winning strategy for player I in G_A; there are infinitely many others.

This section includes two theorems about the determinacy of games. The first, Theorem 5.26, says that some games are undetermined. It will be apparent from our proof that there are $2^{\aleph_0}$ many strategies and $2^{2^{\aleph_0}}$ many games. In particular, there are strictly more games than strategies. By itself, this is not an argument that there are undetermined games because some strategies win in more than one game. For example, if σ is a winning strategy for player I in G_A and $A \subseteq B$, then σ is also a winning strategy for player I in G_B. So more work than mere counting is needed to see that there are undetermined games.

Theorem 5.26 (Gale–Stewart) *There is an undetermined subset of $^\omega\omega$.*

Proof First we claim that if σ is a strategy, then the two functions

$$b \mapsto \sigma * b$$

and

$$a \mapsto a * \sigma$$

are injections from $^\omega\omega$ to itself. This claim is clear because b is the sequence of odd values of $\sigma * b$ and a is the sequence of even values of $a * \sigma$.

Next observe that

$$|^\omega\omega| = \aleph_0^{\aleph_0} = 2^{\aleph_0}$$

and

$$|\{\sigma \mid \sigma \text{ is a strategy}\}| = \left|^{(<^\omega\omega)}\omega\right| = \aleph_0^{(\aleph_0^{<\aleph_0})} = \aleph_0^{\aleph_0} = 2^{\aleph_0}.$$

Say

$$\{\sigma \mid \sigma \text{ is a strategy}\} = \left\{\sigma_\alpha \mid \alpha < 2^{\aleph_0}\right\}.$$

Now choose a_θ and b_θ by recursion on $\theta < 2^{\aleph_0}$ as follows. Assume that a_η and b_η have been selected for $\eta < \theta$. Let

$$A_\theta = \{a_\eta * \sigma_\eta \mid \eta < \theta\}$$

and

$$B_\theta = \{\sigma_\eta * b_\eta \mid \eta < \theta\}.$$

Notice that

$$|A_\theta| \le \theta < 2^{\aleph_0}.$$

Because

$$b \mapsto \sigma_\theta * b$$

is an injection, we can pick $b_\theta \in {}^\omega\omega$ such that

$$\sigma_\theta * b_\theta \notin A_\theta.$$

This determines

$$B_{\theta+1} = \{\sigma_\eta * b_\eta \mid \eta \le \theta\}.$$

Notice that

$$|B_{\theta+1}| \leq \theta + 1 < 2^{\aleph_0}.$$

Because

$$a \mapsto a * \sigma_\theta$$

is an injection, we can pick $a_\theta \in {}^\omega\omega$ such that

$$a_\theta * \sigma_\theta \notin B_{\theta+1}.$$

That completes the definition of $\langle a_\theta \mid \theta < 2^{\aleph_0} \rangle$ and $\langle b_\theta \mid \theta < 2^{\aleph_0} \rangle$.
Let

$$A = \{a_\eta * \sigma_\eta \mid \eta < 2^{\aleph_0}\}$$

and

$$B = \{\sigma_\eta * b_\eta \mid \eta < 2^{\aleph_0}\}.$$

From the recursive definition, it is clear that

$$A \cap B = \emptyset.$$

Let σ be an arbitrary strategy. Say $\sigma = \sigma_\theta$. Then σ is not a winning strategy for player II in G_A because

$$a_\theta * \sigma_\theta \in A.$$

And σ is not a winning strategy for player I in G_A because

$$\sigma_\theta * b_\theta \in B.$$

Therefore, A is undetermined. $\square$

It is worth noting how the Axiom of Choice was used in the previous proof. Without it, we would not necessarily be able to index all the strategies with ordinals at the start. The proof was yet another example of a diagonal argument. Intuitively, we diagonalized over all strategies to make sure that none of them work, handling the θth strategy at stage θ.

We have seen that some games are undetermined. But many of the games that come up in practice, open games for example, turn out to be determined. This is more important because it has mathematical applications.

Theorem 5.27 (Gale–Stewart) *Every open subset of ${}^\omega\omega$ is determined.*

Proof Let U be an open subset of $^\omega\omega$. Suppose that player I does not have a winning strategy in G_U. We must show that player II has a winning strategy in G_U. Let

$$C = {}^\omega\omega - U.$$

Then C is closed, so by Lemma 5.16, there is a tree T on ω such that $C = [T]$. For $s \in {}^{<\omega}\omega$, let

$$U_s = \{x \mid s^\frown x \in U\}.$$

Notice that

$$U_{\langle\,\rangle} = U.$$

Let W be the set

$$\{s \in {}^{<\omega}\omega \mid \text{dom}(s) \text{ is even and I has a winning strategy in } G_{U_s}\}.$$

We refer to the elements of W as *winning positions for player I in* G_U. The following conditions are obviously true.

1. $\langle\,\rangle \notin W$.
2. Let $s \in {}^{<\omega}\omega$ such that $\text{dom}(s)$ is even but $s \notin W$. Then:
   (a) For every $k < \omega$, there is $\ell < \omega$ such that $s^\frown\langle k, \ell\rangle \notin W$.
   (b) There exists $x \in C$ such that $x \restriction \text{dom}(s) = s$.

From conditions (1) and (2a) we can read off a certain strategy

$$\tau : {}^{<\omega}\omega \to \omega$$

such that, for all $a \in {}^\omega\omega$ and $n < \omega$,

$$(a * \tau) \restriction 2n \notin W.$$

Just to be specific, given $s \in {}^{<\omega}\omega$ such that $\text{dom}(s)$ is even but $s \notin W$, for every $k < \omega$, define

$$\tau(s^\frown\langle k\rangle) = \text{the least } \ell < \omega \text{ such that } s^\frown\langle k, \ell\rangle \notin W.$$

The other values of τ are irrelevant, so make them zero. Sometimes this τ is called a *non-losing* strategy for player II because it avoids winning positions for player I. We claim that τ is a winning strategy for player II in G_U. To see this, let $a \in {}^\omega\omega$ and $y = a * \tau$. We must show that $y \notin U$. By condition (2b), for every $n < \omega$, there exists $x_n \in C$ such that

$$x_n \restriction 2n = y \restriction 2n.$$

This can be expressed by the inequality

$$d(x_n, y) \leq 1/2^{2n}.$$

Hence

$$\lim_{n \to \infty} x_n = y.$$

Since C is closed,

$$y \in C = {}^\omega\omega - U.$$

A more direct way to argue this last part is to note that, for every $n < \omega$,

$$y \restriction 2n = x_n \restriction 2n \in T$$

hence

$$y \in [T] = C = {}^\omega\omega - U.$$

$\square$

Exercises

Exercise 5.24 Use Theorem 5.27 to explain why, in chess, either White has a winning strategy, or Black has a strategy to avoid losing.

Exercise 5.25 Prove there is an undetermined set $B \subseteq {}^\omega\omega$ such that ${}^\omega\omega - B$ is determined. *Hint:* By Theorem 5.26, there is an undetermined set A. Do not worry about how A was constructed. Rather, take A as given and define B from A in a way that takes advantage of the asymmetry that player I goes first.

Exercise 5.26 Let C be a closed subset of ${}^\omega\omega$. Prove that C is determined. *Hint:* One approach is to model the proof on that of Theorem 5.27. An easier approach is to use the statement of Theorem 5.27 and derive closed determinacy as a corollary. But the proof is not completely trivial because, by Exercise 5.25, there are determined sets whose complements are not determined.

For the following series of exercises, we define another kind of game, called a *perfect set game* for reasons that will become apparent. For $A \subseteq {}^\omega 2$, let G_A^* be the game whose runs have the following pattern.

I	s_0		s_1		s_2		$\cdots$
II		n_0		n_1		n_2	$\cdots$

At stage $2i$, player I must play $s_i \in {}^{<\omega}2$ or else he loses. At stage $2i + 1$, player II must play $n_i < 2$ or else he loses. At the end of the run, player I wins if

$$s_0 {}^\frown \langle n_0 \rangle {}^\frown s_1 {}^\frown \langle n_1 \rangle {}^\frown s_2 {}^\frown \langle n_2 \rangle \cdots \in A.$$

Otherwise, player II wins the run.

Exercise 5.27 Provide formal definitions of the following terminology.

1. σ is a strategy for player I in G_A^*.
2. σ is a winning strategy for player I in G_A^*.
3. τ is a strategy for player II in G_A^*.
4. τ is a winning strategy for player II in G_A^*.

Exercise 5.28 Let $A \subseteq {}^\omega 2$. Suppose that there exists a perfect tree T on 2 such that $[T] \subseteq A$. Prove that player I has a winning strategy in G_A^*.

Exercise 5.29 Let $A \subseteq {}^\omega 2$. Suppose that player I has a winning strategy in G_A^*. Prove that there exists a perfect tree T on 2 such that $[T] \subseteq A$.

Exercise 5.30 Let $A \subseteq {}^\omega 2$. Suppose that A is countable. Prove that player II has a winning strategy in G_A^*.

Exercise 5.31 This exercise is harder than the previous three but it is the most important. Let $A \subseteq {}^\omega 2$. Suppose that player II has a winning strategy in G_A^*. Prove that A is countable by completing the following outline.

Let τ be a winning strategy for player II in G_A^*. Suppose that p is a position of even length $2j$. If $2j = 0$, then

$$p = \langle \, \rangle,$$

whereas if $2j > 0$, then we may specify that

$$p = \langle s_0, n_0, \ldots, s_{j-1}, n_{j-1} \rangle.$$

Saying that p has even length is the same as saying that, starting from p, it is player I's turn to move. Assume, in addition, that p is *consistent* with τ, by which we mean that

$$n_0 = \tau(\langle s_0 \rangle),$$

$$n_1 = \tau(\langle s_0, n_0, s_1 \rangle),$$

$$n_1 = \tau(\langle s_0, n_0, s_1, n_1, s_2 \rangle),$$

and so on for every n_i with $i < j$. Now define

$$p^* = s_0 {}^\frown \langle n_0 \rangle {}^\frown \ldots {}^\frown s_{j-1} {}^\frown \langle n_{j-1} \rangle.$$

Notice that $p^* \in {}^{<\omega}2$ and the domain of p^* is its finite cardinality

$$|p^*| = j + \sum_{i<j} |s_i|.$$

For $x \in {}^\omega 2$, we say that p *rejects* x iff

- $p^* \subset x$ and
- if q is a position such that
  - q extends p,
  - q has even length (so it is player I's turn to move after q), and
  - q is consistent with τ,

  then $q^* \not\subset x$.

1. Prove that each p as above rejects exactly one $x \in {}^\omega 2$. *Hint*: Define $x(m)$ by recursion on $m < \omega$. Start by setting $x(m) = p^*(m)$ for all $m < |p^*|$. Now suppose that $m \geq |p^*|$. There is a unique $s \in {}^{<\omega}2$ such that

$$x \restriction m = p^* {}^\frown s.$$

   Let

$$n = \tau(p {}^\frown \langle s \rangle).$$

   If $n = 0$, then put $x(m) = 1$. Otherwise $n = 1$, in which case put $x(m) = 0$. Show that p rejects x and, if p rejects y, then $y = x$.

2. Prove that if $x \in A$, then there exists p as above that rejects x. *Hint*: Suppose otherwise and contradict the assumption that τ is a winning strategy for player II.

3. Use the previous results 1 and 2 to conclude that A is countable.
   Hint: Count positions.

Exercise 5.32 Let $C \subseteq {}^\omega 2$ and assume that C is closed. Sketch a proof that G_C^* is determined. In other words, prove that either player I has a winning strategy in G_C^*, or else player II does. *Hint*: All of the ideas are contained in Theorem 5.27 and Exercise 5.26 but writing up a complete proof is challenging because the notation is complicated.

Exercise 5.33 Use Exercises 5.29, 5.31 and 5.32 to prove that if C is a closed subset of ${}^\omega 2$, then either C is countable or C has a perfect subset. Notice this also follows from the Cantor perfect set theorem, which was the subject of Exercise 5.11, but the two proofs are very different.

5.5 Ramsey theory

Ramsey theory is often introduced with the following scenario. Imagine that you are hosting a party and you would like to invite enough people so that either there is a trio of guests who have met the other two in the trio, or there is a trio of guests who have met neither of the other two in the trio. This can be modeled mathematically as follows. Let I represent the set of invited guests and

$$[I]^2 = \{\{a, b\} \mid a, b \in I \text{ and } a \neq b\}$$

be the set of pairs of guests. Let the function

$$F : [I]^2 \to 2$$

be given by

$$F(\{a, b\}) = \begin{cases} 0 & a \text{ and } b \text{ have not met each other} \\ 1 & a \text{ and } b \text{ have met each other.} \end{cases}$$

Such a function is referred to as a *coloring of pairs from I by two colors*. The two *colors* are 0 and 1. So $2 = \{0, 1\}$ is the set of colors. A subset $H \subseteq I$ is called *homogeneous for F* iff there exists $k \in 2$ such that, for all $a, b \in H$, if $a \neq b$, then

$$F(\{a, b\}) = k.$$

Your goal as host is to choose I large enough so that, for every F as above, there is an H as above with $|H| = 3$. By the following two exercises, you should invite at least six guests.

Exercise 5.34 Prove that for every function

$$F : [6]^2 \to 2,$$

there exists $H \subseteq 6$ and $k \in 2$ such that $|H| = 3$ and, for all $a, b \in H$, if $a \neq b$, then

$$F(\{a, b\}) = k.$$

Exercise 5.35 Find a function

$$F : [5]^2 \to 2$$

such that, for every $H \subseteq 5$ and $k \in 2$, if $|H| = 3$, then there exists $a, b \in H$ such that $a \neq b$ and

$$F(\{a, b\}) \neq k.$$

The phenomenon described above has many important extensions. Given $m < \omega$ and a set I, define

$$[I]^m = \{p \subseteq I \mid |p| = m\}.$$

The finite Ramsey theorem says that, for all positive $\ell, m, n < \omega$, there exists $r < \omega$ such that, given

- a set I such that $|I| \geq r$, and
- a function $F : [I]^m \to \ell$,

there exist

- a subset $H \subseteq I$ such that $|H| \geq n$, and
- a number $k < \ell$

with the property that, for every $p \in [H]^m$,

$$F(p) = k.$$

Exercises 5.34 and 5.35 show that, with $m = 2$ (colorings of pairs), $\ell = 2$ (two colors) and $n = 3$ (homogeneous set with three elements), the least witness to the finite Ramsey theorem is $r = 6$. The finite Ramsey theorem is typically proved in a basic course on discrete mathematics; we will not prove it here. The following infinite Ramsey theorem and results like it are of significant importance in set theory and its applications. The way we have organized the proof explains why it sits in our chapter on trees.

Theorem 5.28 (Ramsey) *Let $0 < \ell < \omega$ and F be a function of the form*

$$F : [\omega]^2 \to \ell.$$

Then there exists $k < \ell$ and an infinite $H \subseteq \omega$ such that, for every $p \in [H]^2$,

$$F(p) = k.$$

Proof Let T be the set of strictly increasing $s \in {}^{<\omega}\omega$ such that, for every $m < \mathrm{dom}(s)$, there exists $k < \ell$ such that, for every $n < \mathrm{dom}(s)$, if $m < n$, then

$$F(\{s(m), s(n)\}) = k.$$

Then T is a tree on ω. In this context, *strictly increasing* means that, for all $m < n < \mathrm{dom}(s)$,

$$s(m) < s(n).$$

We will prove that T has an infinite branch. But first let us show why the existence of such a branch suffices to prove the theorem. Suppose that x is an infinite branch of T. For each $m < \omega$, let k_m be the unique $k < \ell$ such that, for every $n < \omega$, if $m < n$, then

$$F(\{x(m), x(n)\}) = k.$$

Since $\{k_m \mid m < \omega\} \subseteq \ell$ is finite, there exists an infinite $S \subseteq \omega$ and $k < \ell$ such that, for every $m \in S$,

$$k_m = k.$$

Let

$$H = \{x(m) \mid m \in S\}.$$

Since S is infinite and x is increasing, H is infinite. Moreover, for every $a, b \in H$,

$$F(\{a, b\}) = k.$$

Thus H witnesses the conclusion of the theorem.

Now we construct an infinite branch x through T. By recursion on $n < \omega$, we define $x(n)$ and, simultaneously, $k_n < \ell$ and an infinite $I_n \subseteq \omega$. Start by defining

$$I_0 = \omega$$

and

$$x(0) = 0.$$

Now assume we are given I_n and it is an infinite subset of ω. Let

$$x(n) = \min(I_n).$$

This is consistent with our having set $x(0) = 0$. For each $k < \ell$, let

$$J_k = \{i \in I_n \mid i > x(n) \text{ and } F(\{x(n), i\}) = k\}.$$

Then

$$I_n - \{x(n)\} = J_0 \cup \cdots \cup J_{\ell-1}.$$

Since $I_n - \{x(n)\}$ is infinite, there exists $k < \ell$ such that J_k is infinite. Let k_n be the least such k and

$$I_{n+1} = J_{k_n}.$$

That completes the recursive construction. By induction on $n < \omega$, it is obvious that

$$x(n) = \min(I_n)$$

and, for every $m < n$,

$$I_n \subseteq I_{m+1} \subseteq I_m - \{x(m)\},$$

$$x(m) < x(n)$$

and

$$F(\{x(m), x(n)\}) = k_m.$$

Therefore, x is an infinite branch through T. $\square$

Exercise 5.36 Let $0 < m < \omega$. Show that Theorem 5.28 remains true if "2" is replaced by "m" in its statement. *Hint*: Use induction on m. The case $m = 1$ follows from the *pigeonhole principle*. (If you partition an infinite set into finitely many pieces, then one of the pieces must be infinite.) Think of the proof of Theorem 5.28 as showing that the case $m = 1$ implies the case $m = 2$. Generalize this to see that case m implies case $m + 1$.

Exercise 5.37 Prove that Theorem 5.28 becomes false if "ω" is replaced by "ω_1" and "infinite" is replaced by "uncountable" using the following example. Fix an injection

$$g : \omega_1 \to \mathbb{R}.$$

In order to avoid possible confusion, we write $<_\mathbb{R}$ for the usual order of $\mathbb{R}$ here. Define

$$F : [\omega_1]^2 \to 2$$

by

$$F(\{\alpha, \beta\}) = \begin{cases} 0 & \text{if } \alpha < \beta \text{ and } g(\alpha) >_\mathbb{R} g(\beta) \\ 1 & \text{if } \alpha < \beta \text{ and } g(\alpha) <_\mathbb{R} g(\beta). \end{cases}$$

Prove that there is no uncountable set that is homogeneous for F. In other words, prove that if H is an uncountable subset of ω_1 and $k < 2$, then there exists $p \in [H]^2$ such that $F(p) = k$. *Hint*: Argue by contradiction and use the fact that between any two real numbers there is a rational number.

5.6 Trees of uncountable height

So far, all the trees we have looked at have at most ω many levels. This section is on trees of height ω_1. We might expect that our results about trees of height ω would lift to theorems about trees of height ω_1. For example, recall that Theorem 5.17 says that if T is a subtree of $^{<\omega}\omega$ and, for every $n < \omega$,

$$0 < |T \cap {}^n\omega| < \aleph_0,$$

then T has an ω-branch, i.e., there exists $b : \omega \to \omega$ such that, for every $n < \omega$,

$$b \upharpoonright n \in T.$$

Does this statement remain true if we replace ω by ω_1 and $\aleph_0$ by $\aleph_1$? It turns out that the answer is no; counterexamples are called *Aronszajn* trees and the point of this section is to construct one. This is a big topic; in order to keep this section manageable, we still do not give the most general definition of *tree*.

Definition 5.29   A *subtree* of $^{<\omega_1}\omega$ is a subset $T \subseteq {}^{<\omega_1}\omega$ such that, for all $s \in T$ and $\alpha < \text{dom}(s)$,

$$s \restriction \alpha \in T.$$

An ω_1-*branch* of T is a function $b : \omega_1 \to \omega$ such that, for every $\alpha < \omega_1$,

$$b \restriction \alpha \in T.$$

Theorem 5.30 (Aronszajn)   *There exists a subtree T of* $^{<\omega_1}\omega$ *such that*

$$0 < |T \cap {}^{\alpha}\omega| < \aleph_1$$

for every $\alpha < \omega_1$ but T has no ω_1-branch.

Proof   Let

$$I = \{s \in {}^{<\omega_1}\omega \mid s \text{ is an injection}\}.$$

The good news is that I is obviously a subtree of $^{<\omega_1}\omega$ and I has no ω_1-branch because there is no injection from ω_1 to ω. The bad news is that, whenever $\omega \le \alpha < \omega_1$,

$$|I \cap {}^{\alpha}\omega| = |\{s \in {}^{\alpha}\omega \mid s \text{ is an injection}\}| = 2^{\aleph_0} \ge \aleph_1,$$

which is too large. We will find an appropriate subtree $T \subseteq I$. For $\beta < \omega_1$ and $s, t \in {}^{\beta}\omega$, define

$$s \sim_\beta t \iff |\{\alpha < \beta \mid s(\alpha) \ne t(\alpha)\}| < \aleph_0.$$

It is easy to see that $\sim_\beta$ is an equivalence relation on $^{\beta}\omega$.

Claim 5.30.1   *There is a sequence $\langle s_\alpha \mid \alpha < \omega_1 \rangle$ so that, for every $\beta < \omega_1$,*

$$s_\beta \in I \cap {}^{\beta}\omega$$

and, for every $\alpha < \beta$,

$$s_\alpha \sim_\alpha s_\beta \restriction \alpha.$$

Assuming Claim 5.30.1, if we define

$$T = \bigcup_{\alpha < \omega_1} \{t \in I \cap {}^{\alpha}\omega \mid t \sim_\alpha s_\alpha\}$$

then T witnesses the statement of the theorem. To see this, first observe that T is a subtree of $^{<\omega_1}\omega$. The point is that T is closed downward under restriction. This is true because if $\alpha < \beta < \omega_1$ and

$$t \in T \cap {}^\beta \omega,$$

then

$$t \sim_\beta s_\beta,$$

hence

$$t \upharpoonright \alpha \sim_\alpha s_\beta \upharpoonright \alpha \sim_\alpha s_\alpha,$$

so

$$t \upharpoonright \alpha \in T \cap {}^\alpha \omega.$$

Second, observe that the levels of T are countable. This is because, for every $\gamma < \omega_1$ and

$$t \in T \cap {}^\gamma \omega,$$

there are $n < \omega$ and $\alpha_0 < \cdots < \alpha_{n-1} < \gamma$ such that $t(\beta) = s_\gamma(\beta)$ for every $\beta < \gamma$ except if $\beta = \alpha_m$ for some $m < n$. Since there are at most countably many ways to make these sorts of finite changes to s_γ, we have that, for every $\gamma < \omega_1$,

$$1 \le |T \cap {}^\gamma \omega| \le \aleph_0.$$

It remains to prove Claim 5.30.1. We define s_β by recursion on $\beta < \omega_1$. In addition to maintaining the two requirements of the claim, we also maintain the *extra property* that, for every $\beta < \omega_1$,

$$|\omega - \mathrm{ran}(s_\beta)| = \aleph_0.$$

This is so that the recursion does not run out of steam, as you will see. We have no choice but to set $s_0 = \emptyset$. If s_α has been defined, then pick $k \in \omega - \mathrm{ran}(s_\beta)$ and define

$$s_{\alpha+1} = s_\alpha \cup \{(\alpha, k)\}.$$

Now suppose that $\gamma < \omega_1$ is a limit ordinal and we have

$$\langle s_\alpha \mid \alpha < \gamma \rangle$$

satisfying the three requirements. The definition of s_γ is somewhat tricky in this case. Since $\mathrm{cf}(\gamma) = \omega$, there are ordinals

$$\beta_0 < \beta_1 < \cdots \beta_i < \cdots < \gamma = \sup_{i<\omega} \beta_i.$$

By recursion on $i < \omega$, define a sequence $\langle t_i \mid i < \omega \rangle$ from I such that, for every $j < \omega$,

$$t_j \sim_{\beta_j} s_{\beta_j}$$

and, for every $i < j$,

$$t_i = t_j \upharpoonright \beta_i.$$

Start with

$$t_0 = s_{\beta_0}.$$

Suppose we are given t_i with

$$t_i \sim_{\beta_i} s_{\beta_i}.$$

Let us say that a pair (α, α') is *bad* iff

$$\alpha < \beta_i \leq \alpha' < \beta_{i+1}$$

and

$$t_i(\alpha) = s_{\beta_{i+1}}(\alpha').$$

The reason we call them *bad* is because their existence prevents us from setting

$$t_{i+1}(\eta) = \begin{cases} t_i(\eta) & \text{if } \eta < \beta_i \\ s_{\beta_{i+1}}(\eta) & \text{if } \eta \in \beta_{i+1} - \beta_i. \end{cases}$$

Remember that t_{i+1} is supposed to be an injection! However, we claim that there are only finitely many bad pairs. First note that if (α, α') is a bad pair, then

$$t_i(\alpha) \neq s_{\beta_{i+1}}(\alpha).$$

This is because $s_{\beta_{i+1}}$ is an injection, so it cannot take on the same value, $t_i(\alpha)$, twice. From this and the fact that

$$t_i \sim_{\beta_i} s_{\beta_{i+1}} \upharpoonright \beta_i,$$

it follows that there are only finitely many first coordinates of bad pairs. Finally, observe that if (α, α') and (α, α'') are bad pairs, then

$$s_{\beta_{i+1}}(\alpha') = t_i(\alpha) = s_{\beta_{i+1}}(\alpha''),$$

so $\alpha' = \alpha''$, again because $s_{\beta_{i+1}}$ is an injection. This shows that there are only finitely many bad pairs. Say the number of bad pairs is n where $n < \omega$. List the bad pairs as (α_m, α'_m) for $m < n$. Then pick distinct

$$k_m \in \omega - (\text{ran}(t_i) \cup \text{ran}(s_{\beta_{i+1}}))$$

for $m < n$. This is possible by the *extra property* on s_{β_i} and $s_{\beta_{i+1}}$, and the fact that $t_i \sim_{\beta_i} s_{\beta_i} \sim_{\beta_i} s_{\beta_{i+1}} \upharpoonright \beta_i$. Our solution to the problem of bad pairs is to define

$$t_{i+1}(\eta) = \begin{cases} t_i(\eta) & \text{if } \eta < \beta_i \\ s_{\beta_{i+1}}(\eta) & \text{if } \eta \in \beta_{i+1} - \beta_i \text{ and } \eta \neq \alpha'_m \text{ for every } m < n \\ k_m & \text{if } \eta = \alpha'_m. \end{cases}$$

Obviously, this satisfies our requirements for t_{i+1}. Now, having completed the definition of $\langle t_i \mid i < \omega \rangle$, we put

$$t = \bigcup_{i<\omega} t_i.$$

The good news is that t meets the first and second requirements for s_γ of Claim 5.30.1, namely that

$$t \in I \cap {}^\gamma \omega$$

and, for every $\alpha < \gamma$,

$$s_\alpha \sim_\alpha t \upharpoonright \alpha.$$

The problem is that it might not satisfy the *extra property* that

$$|\omega - \text{ran}(t)| = \aleph_0.$$

To fix up this potential problem, put

$$s_\gamma(\alpha) = \begin{cases} t(\beta_{2i}) & \text{if } \alpha = \beta_i \text{ for some } i < \omega \\ t(\alpha) & \text{if } \alpha \in \gamma - \{\beta_i \mid i < \omega\}. \end{cases}$$

Then all three requirements on s_γ are met. $\qquad\square$

Exercises

Exercise 5.38 Let T be the tree constructed in the proof of Theorem 5.30. Suppose that $\alpha < \beta < \omega_1$ and $u \in T \cap {}^\alpha \omega$. Explain why there exists $v \in T \cap {}^\beta \omega$ such that $u = v \upharpoonright \alpha$. We say that u has *extensions to every level of* T.

Exercise 5.39 Let T be the tree constructed in the proof of Theorem 5.30. Suppose that $\alpha < \omega_1$ and $u \in T \cap {}^\alpha\omega$. Explain why there exist $\beta > \alpha$ and $v, w \in T \cap {}^\beta\omega$ such that

$$u = v \upharpoonright \alpha = w \upharpoonright \alpha$$

but

$$v \neq w.$$

We say that u *splits* in T.

Exercise 5.40 Let T be the tree constructed in the proof of Theorem 5.30. Find a sequence $\langle u_\alpha \mid \alpha < \omega_1 \rangle$ of members of T such that, for all $\alpha < \beta < \omega_1$,

$$u_\alpha \not\subseteq u_\beta$$

and

$$u_\beta \not\subseteq u_\alpha.$$

In this case, we say that $\langle u_\alpha \mid \alpha < \omega_1 \rangle$ is an *antichain* of T because u_α and u_β are *incomparable* (cannot be compared using $\subseteq$) whenever $\alpha \neq \beta$.

Exercise 5.41 Let T be a subtree of ${}^{<\omega_1}\omega$ with an ω_1-branch

$$b : \omega_1 \to \omega.$$

Assume that every $u \in T$ splits in T. Prove that T has an uncountable antichain.

6
Dense linear orderings

This chapter is mainly about two theorems, one due to Cantor, the other to Dedekind, which are characterizations of the rationals, $\mathbb{Q}$, and the reals, $\mathbb{R}$, in terms of their respective orderings.

6.1 Definitions and examples

Recall that $(A, \prec)$ is a strict linear ordering iff $\prec$ is a transitive, irreflexive and total relation on A. As usual, we write $x \preccurlyeq y$ for $x \prec y$ or $x = y$.

Definition 6.1 $(A, \prec)$ is a *dense linear ordering* iff $(A, \prec)$ is a strict linear order with at least two elements and, for all $x, y \in A$, if $x \prec y$, then there exists $z \in A$ such that $x \prec z \prec y$.

We required A to have at least two elements because, otherwise, $(0, <)$ and $(1, <)$ would be dense linear orderings, which would be counterintuitive. It follows easily from Definition 6.1 that every dense linear ordering is infinite.

Definition 6.2 If $(A, \prec)$ is a strict linear ordering and $L, R \in A$, then we say:

- L is a *left endpoint* of $(A, \prec)$ iff $L \preccurlyeq x$ for every $x \in A$.
- R is a *right endpoint* of $(A, \prec)$ iff $x \preccurlyeq R$ for every $x \in A$.

In this chapter we are mainly interested in dense linear orderings without endpoints. Here are several examples, each referring to the usual ordering of real numbers.

- $\mathbb{Q} = \{m/n \mid m, n \in \mathbb{Z} \text{ and } n \neq 0\}$

- $\mathbb{R}$ = the set of real numbers
- the open interval $(0,1)$
- $(0,1) \cap \mathbb{Q}$

The reader has not yet seen definitions of $\mathbb{Q}$ and $\mathbb{R}$ in this book but has intuition about these based on doing mathematics since childhood. Temporarily, we rely only on that intuition.

Continuing with our introduction, we make the following key definition, which is really just a repetition of Definition 3.27.

Definition 6.3 We say that f *is an isomorphism from* $(A, \prec_A)$ *to* $(B, \prec_B)$ and write

$$f : (A, \prec_A) \simeq (B, \prec_B)$$

iff f is a bijection from A to B and, for all $x, y \in A$,

$$x \prec_A y \iff f(x) \prec_B f(y).$$

We say that $(A, \prec_A)$ *is isomorphic to* $(B, \prec_B)$ and write

$$(A, \prec_A) \simeq (B, \prec_B)$$

iff there is an isomorphism $f : (A, \prec_A) \simeq (B, \prec_B)$.

For example, with the usual ordering on the real line, the open interval $(-\pi/2, \pi/2)$ is isomorphic to $\mathbb{R}$ as witnessed by the function

$$x \mapsto \tan(x).$$

So,

$$(-\pi/2, \pi/2) \simeq \mathbb{R}.$$

It also turns out that

$$(-\pi/2, \pi/2) \cap \mathbb{Q} \simeq \mathbb{Q}$$

but since $\arctan(1) = \pi/4 \notin \mathbb{Q}$, a function other than tangent is needed. We will see why there is such an isomorphism later.

As an example of non-isomorphism, observe that $\mathbb{Q} \not\simeq \mathbb{R}$ because

$$|\mathbb{Q}| = \aleph_0 \neq 2^{\aleph_0} = |\mathbb{R}|,$$

so there is not even a bijection between $\mathbb{Q}$ and $\mathbb{R}$. A more subtle example is the fact that

$$\mathbb{R} \smallfrown \mathbb{Q} \not\simeq \mathbb{R}.$$

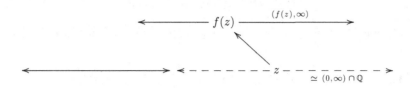

Figure 6.1 $\mathbb{R}^\frown\mathbb{Q} \not\simeq \mathbb{R}$

Here, $\mathbb{R}^\frown\mathbb{Q}$ is the concatenation of $\mathbb{R}$ followed by $\mathbb{Q}$. It is easy to see that $\mathbb{R}^\frown\mathbb{Q}$ is a dense linear ordering without endpoints. Suppose for contradiction that

$$f : \mathbb{R}^\frown\mathbb{Q} \simeq \mathbb{R}.$$

Let z be the zero of $\mathbb{Q}$ in the concatenation $\mathbb{R}^\frown\mathbb{Q}$.[1] Then, $f(z) \in \mathbb{R}$ and

$$(0, \infty) \cap \mathbb{Q} \simeq (f(z), \infty).$$

The set on the left has cardinality $\aleph_0$ and the set on the right has cardinality $2^{\aleph_0}$, so there is no bijection between them. Figure 6.1 illustrates the point.

This brings up an interesting question: what about $\mathbb{Q}$ versus $\mathbb{Q}^\frown\mathbb{Q}$? In the next section, we will state and prove Theorem 6.5, which implies that

$$\mathbb{Q} \simeq \mathbb{Q}^\frown\mathbb{Q}.$$

Theorem 6.5 also implies that

$$(-\pi/2, \pi/2) \cap \mathbb{Q} \simeq \mathbb{Q},$$

which we already mentioned.

---

[1] Formally, the underlying set of the strict linear ordering $\mathbb{R}^\frown\mathbb{Q}$ is $(\{0\} \times \mathbb{R}) \cup (\{1\} \times \mathbb{Q})$ and z is the ordered pair $(1, 0)$.

6.2 Rational numbers

This section has two theorems, the first of which should come as
no surprise.

Theorem 6.4   *There is a countable dense linear ordering without
endpoints.*

Exercise 6.1 outlines a proof of Theorem 6.4 based only on what
we already know about natural numbers, their ordering and their
arithmetic. Obviously, $(\omega, <)$ is not dense and has a left endpoint,
so some work is needed. It would be tempting to blurt out that
$\mathbb{Q}$ with its ordering already witnesses Theorem 6.4 but this would
be cheating since the point here is to say what $\mathbb{Q}$ is up to isomorphism.

The next theorem tells us that in a certain sense it does not
matter which countable dense linear ordering without endpoints
we work with since they are all the same up to isomorphism. It
says we can take our pick and make it $\mathbb{Q}$. The technique used
in the proof, a *back-and-forth* construction, is also important in
other parts of mathematics. An application of this technique to
the theory of Boolean algebras is the topic of Exercise 6.6.

Theorem 6.5 (Cantor)   *Let $(A, \prec_A)$ and $(B, \prec_B)$ be countable
dense linear orderings without endpoints. Then*

$$(A, \prec_A) \simeq (B, \prec_B).$$

Proof   Say $A = \{a_0, a_1, \dots\}$ and $B = \{b_0, b_1, \dots\}$. Warning! It is
definitely not true that $i < j \iff a_i \prec_A a_j$ because $(\omega, <)$ is a
wellordering whereas $(A, \prec_A)$ is an illfounded relation.

By recursion on $n < \omega$, we will define finite bijections

$$f_n : \mathrm{dom}(f_n) \to \mathrm{ran}(f_n)$$

such that

$$\{a_i \mid i < n\} \subseteq \mathrm{dom}(f_n) \subseteq A$$

and

$$\{b_i \mid i < n\} \subseteq \mathrm{ran}(f_n) \subseteq B$$

and

$$f_n : (\mathrm{dom}(f_n), \prec_A) \simeq (\mathrm{ran}(f_n), \prec_B).$$

We also maintain that, for all $m < n < \omega$,

$$f_m \subseteq f_n.$$

In other words,

$$f_n \restriction \mathrm{dom}(f_m) = f_m.$$

Mere success in this construction is enough because if

$$f = \bigcup_{n < \omega} f_n,$$

then it is easy to verify that

$$f : (A, \prec_A) \simeq (B, \prec_B).$$

For example, to verify that f defined this way is order preserving, note that if

$$i, j < n < \omega,$$

then

$$a_i \prec_A a_j \iff f_n(a_i) \prec_B f_n(a_j) \iff f(a_i) \prec_B f(a_j)$$

because $a_i, a_j \in \mathrm{dom}(f_n)$, f_n is order preserving and

$$f_n = f \restriction \mathrm{dom}(f_n).$$

Here is the recursive definition of f_n. For the base step, set $f_0 = \emptyset$. Now assume that f_n has been defined. The definition of f_{n+1} is a two-step process: back and forth.

Forth step *We define a finite isomorphism*

$$g : (\mathrm{dom}(g), \prec_A) \simeq (\mathrm{ran}(g), \prec_B).$$

with

$$\mathrm{dom}(g) = f_n \cup \{a_n\}.$$

Case 1 $a_n \in \mathrm{dom}(f_n)$.

Set $g = f_n$.

Case 2 $a_n \notin \mathrm{dom}(f_n)$.

We claim that there exists $i < \omega$ such that if

$$g = f_n \cup \{(a_n, b_i)\},$$

then

$$g : (\mathrm{dom}(g), \prec_A) \simeq (\mathrm{ran}(g), \prec_B).$$

The proof of the claim breaks up into three subcases.

Subcase 1 $a_n \prec_A x$ *for every $x \in \mathrm{dom}(f_n)$.*

Since $(B, \prec_B)$ does not have a left endpoint, we may pick $i < \omega$ such that $b_i \prec_B y$ for every $y \in \mathrm{ran}(f_n)$.

Subcase 2 $x \prec_A a_n$ *for every $x \in \mathrm{dom}(f_n)$.*

Since $(B, \prec_B)$ does not have a right endpoint, we may pick $i < \omega$ such that $y \prec_B b_i$ for every $y \in \mathrm{ran}(f_n)$.

Subcase 3   *Otherwise.*

Since $\mathrm{dom}(f_n)$ is finite, we can pick $\ell, r \in \mathrm{dom}(f_n)$ such that

$$\ell \prec_A a_n \prec_A r$$

and, for every $a \in \mathrm{dom}(f_n)$, either

$$a \preccurlyeq_A \ell$$

or

$$r \preccurlyeq_A a.$$

Then,

$$f_n(\ell) \prec_B f_n(r)$$

and, for every $b \in \mathrm{ran}(f_n)$, either

$$b \preccurlyeq_B f_n(\ell)$$

or

$$f_n(r) \preccurlyeq_B b.$$

Because $(B, \prec_B)$ is a dense linear ordering, we can pick $i < \omega$ such that

$$f_n(\ell) \prec_B b_i \prec_B f_n(r).$$

In each of the three subcases, it is clear that the prescribed choice of i works, so the claim is proved, case 2 has been handled and the forth step is complete.

Back step   *We define a finite isomorphism*

$$h : (\mathrm{dom}(h), \prec_A) \simeq (\mathrm{ran}(h), \prec_B).$$

with

$$\mathrm{ran}(h) = \mathrm{ran}(g) \cup \{b_n\}.$$

The process for defining h from g is like the process of defining g from f_n except that we reverse the roles of A and B.

Finally, having completed both the back and forth steps, let $f_{n+1} = h$.   □

6.3 Real numbers

Having characterized $\mathbb{Q}$ with its ordering up to isomorphism, we turn our attention to $\mathbb{R}$. What is $\mathbb{R}$ and which criteria characterize $\mathbb{R}$ with its ordering up to isomorphism? As motivation, consider the set

$$S = \{x \in \mathbb{Q} \mid x^2 < 2\}.$$

Notice that S is bounded above by the rational number $3/2$. This is because, if x is a rational number and $x^2 < 2$, then $x < 3/2$. (If $x \geq 3/2$, then $x^2 \geq 9/4 > 2$.) Based on the mathematics you knew before starting to read this book, you would have said that $\sqrt{2}$ is also an upper bound for S, in fact, $\sqrt{2}$ is the least upper bound for S, which we write

$$\mathrm{lub}(S) = \sqrt{2}.$$

This is still correct except that $\sqrt{2} \notin \mathbb{Q}$ and we have not defined $\mathbb{R}$ so it is not fair to mention $\sqrt{2}$ yet. The moral of this paragraph is that, in passing from $\mathbb{Q}$ to $\mathbb{R}$, we want to include least upper bounds for all bounded sets.

Definition 6.6   Let $(A, \prec_A)$ be a strict linear ordering. If $S \subseteq A$ and $y \in A$, then y is an *upper bound* for S iff $x \preccurlyeq_A y$ for every $x \in S$.

To be clear, $x \preccurlyeq_A y$ means that either $x \prec_A y$ or $x = y$.

Definition 6.7   Let $(A, \prec)$ be a strict linear ordering. If $S \subseteq A$ and $y \in A$, then y is a *least upper bound* for S iff y is an upper bound for S and, for every upper bound z for S, $y \preccurlyeq_A z$.

It is easy to see that S has at most one least upper bound in $(A, \prec_A)$. For if it had two, y and z, then $y \preccurlyeq_A z$ and $z \preccurlyeq_A y$ so $y = z$.

Now we arrive at the property that distinguishes $\mathbb{R}$ from $\mathbb{Q}$.

Definition 6.8   A strict linear ordering $(A, \prec_A)$ has the *least upper bound property* iff for every non-empty $S \subseteq A$, if S has an upper bound in $(A, \prec_A)$, then S has a least upper bound in $(A, \prec_A)$.

We indicated above that $\mathbb{Q}$ with its usual ordering does not have the least upper bound property since

$$\mathrm{lub}(\{x \in \mathbb{Q} \mid x^2 < 2\}) \notin \mathbb{Q}.$$

Of course, this is not the only example of a subset of $\mathbb{Q}$ without a least upper bound in $\mathbb{Q}$. By contrast, the least upper bound property is one of the key properties that characterize $\mathbb{R}$ with its usual ordering.

Another essential property about the usual ordering of $\mathbb{R}$ is that $\mathbb{Q} \subseteq \mathbb{R}$ and between any two real numbers there is a rational number. In other words, $\mathbb{Q}$ is dense in $\mathbb{R}$.

Definition 6.9   If $(B, \prec_B)$ is a strict linear ordering and $A \subseteq B$, then A is *dense in* $(B, \prec_B)$ iff for all $x, y \in B$, there exists $z \in A$ such that $x \prec_B z \prec_B y$.

Theorem 6.5 tells us that in a certain sense it does not matter which countable dense linear ordering without endpoints we take to be $\mathbb{Q}$ since they are all isomorphic. The next theorem tells us that, once we settle on a choice of $\mathbb{Q}$, there is a way to extend $\mathbb{Q}$ to obtain an appropriate choice for $\mathbb{R}$, and this choice for $\mathbb{R}$ is unique up to isomorphisms that fix $\mathbb{Q}$.

Theorem 6.10 (Dedekind)   *Let $(A, \prec_A)$ be a countable dense linear ordering without endpoints. Then there exists $(B, \prec_B)$ such that:*

- *$(B, \prec_B)$ is a dense linear ordering without endpoints;*
- *$(A, \prec_A)$ is a suborder of $(B, \prec_B)$, that is, $A \subseteq B$ and, for all $x, y \in A$,*

$$x \prec_A y \iff x \prec_B y;$$

- *A is dense in $(B, \prec_B)$;*

- $(B, \prec_B)$ has the least upper bound property.

Moreover, if $(B', \prec_{B'})$ has the same four properties as $(B, \prec_B)$, then there is an isomorphism

$$f : (B, \prec_B) \simeq (B', \prec_{B'})$$

such that, for every $x \in A$,

$$f(x) = x.$$

Proof The *moreover* part of Theorem 6.10 is proved by observing that, for every $y \in B$,

$$y = \text{the lub in } (B, \prec_B) \text{ of } \{x \in A \mid x \prec_B y\}$$

and setting

$$f(y) = \text{the lub in } (B', \prec_{B'}) \text{ of } \{x \in A \mid x \prec_B y\}.$$

The details are left to the reader. See Exercise 6.4.

Now we prove the first part of Theorem 6.10. The commutative diagram in Figure 6.2 summarizes a big part of our plan. First, we will define $(D, \prec_D)$ and prove it is a dense linear ordering without endpoints that has the least upper bound property. Then, we will find an isomorphic copy $(C, \prec_C)$ of $(A, \prec_A)$ sitting densely inside of $(D, \prec_D)$. Finally, we will find $(B, \prec_B)$ sitting above $(A, \prec_A)$ the way that $(D, \prec_D)$ sits above $(C, \prec_C)$. What exactly the linear orderings and isomorphisms in the diagram are will be explained soon.

The proof uses the following key definition. We say that a set L is a *left-cut* iff

- $L \subseteq A$,
- $L \neq \emptyset$,
- $L \neq A$,
- for every $x \in A$ and $y \in L$, if $x \prec_A y$, then $x \in L$, and
- for every $x \in L$, there exists $y \in L$ such that $x \prec_A y$.

The next to last condition says that left-cuts are closed to the left. The last condition says that left-cuts do not have a right endpoint.

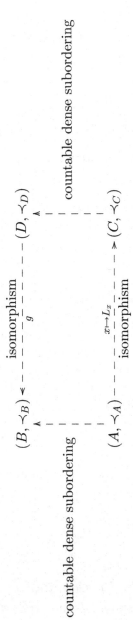

Figure 6.2 *Plan for the proof of Theorem 6.10*

Before continuing, we give a couple of examples of left-cuts in the most interesting case where $(A, \prec_A) = (\mathbb{Q}, <_{\mathbb{Q}})$. Both

$$\{x \in \mathbb{Q} \mid x < 2\}$$

and

$$\{x \in \mathbb{Q} \mid x < 0 \text{ or } x^2 < 2\}$$

are left-cuts of $(\mathbb{Q}, <_{\mathbb{Q}})$. The first example has a least upper bound in $\mathbb{Q}$, namely

$$2 = \mathrm{lub}(\{x \in \mathbb{Q} \mid x < 2\}).$$

But the second example does not have a least upper bound in $\mathbb{Q}$ because $\sqrt{2} \notin \mathbb{Q}$. It is worth keeping these examples in mind in what follows.

Returning to the proof of the theorem, define

$$D = \{L \subseteq A \mid L \text{ is a left-cut}\}$$

and

$$L \prec_D M \iff L \subsetneq M.$$

We claim that $(D, \prec_D)$ is a strict linear ordering. It is obviously transitive and irreflexive, so all that remains is to see that it is total. Suppose that L and M are left-cuts. We must show that $M \subseteq L$ or $L \subsetneq M$. For this, we assume $M \not\subseteq L$ and prove $L \subsetneq M$. Pick $y \in M$ such that $y \notin L$. Consider an arbitrary $x \in L$. Since $(A, \prec_A)$ is a strict linear ordering, either $x \preccurlyeq_A y$ or $y \prec_A x$. If $x \preccurlyeq_A y$, then $x \in M$ since $y \in M$ and M is closed to the left. If $y \prec_A x$, then $y \in L$ since $x \in L$ and L is closed to the left, but this is a contradiction since $y \notin L$. Therefore $x \in M$. Since x was an arbitrary element of L, we have seen that $L \subseteq M$. Clearly, $L \subsetneq M$ since $y \in M - L$. This shows that $(D, \prec_D)$ is total.

Now we verify that $(D, \prec_D)$ has no endpoints. Let $M \in D$. Since M is a left-cut, $M \neq \emptyset$ and $M \neq A$. Pick $y \in M$ and $z \in A - M$. Let $L = \{x \in A \mid x \prec_A y\}$ and $N = \{x \in A \mid x \prec_A z\}$. Routine checking shows that L and N are left-cuts and $L \subsetneq M \subseteq N$. Since $(A, \prec_A)$ has no right endpoint, there exists $z' \in A$ such that $z \prec_A z'$. Let $N' = \{x \in A \mid x \prec_A z'\}$. It is easy to see that $L \subsetneq M \subsetneq N'$. Therefore, M is not an endpoint of $(D, \prec_D)$.

Before proving that $(D, \prec_D)$ has the least upper bound property, we pause again to give a motivating example in the case

$(A, \prec_A) = (\mathbb{Q}, <_\mathbb{Q})$. Let

$$S = \left\{ \{x \in \mathbb{Q} \mid x < y\} \mid y \in \mathbb{Q} \text{ and } y^2 < 2 \right\}.$$

Then S is a non-empty family of left-cuts of $(\mathbb{Q}, <_\mathbb{Q})$. Also, S is bounded in the sense that every element of S is contained in the left-cut $\{z \in \mathbb{Q} \mid z < 3/2\}$. Finally, notice that

$$\bigcup S = \{x \in \mathbb{Q} \mid x < 0 \text{ and } x^2 < 2\}.$$

This example gives a big hint for what is going on.

Back to the proof of the theorem, we are ready to show that $(D, \prec_D)$ has the least upper bound property. Suppose that S is a non-empty subset of D and S has an upper bound in $(D, \prec_D)$. We must prove that S has a least upper bound in $(D, \prec_D)$. Let $M = \bigcup S$. It suffices to show that $M \in D$ and $M = \text{lub}(S)$. Leaving some of the details to the reader, here are the main ideas for showing that M is a left-cut and an upper bound for S.

- If $L \in S$, then $L \subseteq A$. Thus $M \subseteq A$.
- $S \neq \emptyset$ and, if $L \in S$, then $L \neq \emptyset$. Thus $M \neq \emptyset$.
- If $L \in S$, then L has no right endpoint. Therefore M has no right endpoint.
- If $L \in S$, then L is closed to the left. Therefore M is closed to the left.
- If N is an upper bound for S, then $L \subseteq N$ for every $L \in S$, so $M \subseteq N$. Because $N \neq A$, also $M \neq A$.

Now we explain why M is the least upper bound of S. Consider any upper bound N for S. We must show that $M \subseteq N$. Suppose otherwise. Then $N \subsetneq M$ since both are left-cuts. Pick $y \in M - N$. Since $M = \bigcup S$, there exists $L \in S$ such that $y \in L$. Then $L \subseteq N$ because N is an upper bound for S. Thus $y \in N$, which is a contradiction.

Technically, $(A, \prec_A)$ is not a subordering of $(D, \prec_D)$ as we cannot expect $A \subseteq D$. However, we can find a subordering $(C, \prec_C)$ of $(D, \prec_D)$ such that

$$(A, \prec_A) \simeq (C, \prec_C).$$

To see this, for each $y \in A$, let

$$L_y = \{x \in A \mid x \prec_A y\}.$$

Then let

$$C = \{L_y \mid y \in A\}.$$

It is easy to see that the map

$$y \mapsto L_y$$

is an order-preserving injection from $(A, \prec_A)$ to $(D, \prec_D)$ whose range is C.

We must prove that C is dense in $(D, \prec_D)$. Suppose that $K \prec_D M$. Recall this means that $K \subsetneqq M$. Pick $y \in M - K$. Since M is a left-cut, it does not have a largest element, so we may pick $z \in M$ such that $y \prec_A z$. Clearly,

$$K \preccurlyeq_D L_y \prec_D L_z \prec_D M.$$

Observe that L_z is strictly between K and M. This shows that C is dense in $(D, \prec_D)$.

Towards defining $(B, \prec_B)$ as in the statement of Theorem 6.10, first define an injection g with domain D as follows.

- If $x \in A$, then $g(L_x) = x$.
- If $M \in D - C$, then $g(M) = (A, M)$.

In the first case, notice that if $L_x = L_y$, then $x = y$, so g really is a function. In the second case, one point is that $(A, M) \notin A$ because otherwise

$$A \in \{A\} \in \{\{A\}, \{A, M\}\} = (A, M) \in A,$$

which contradicts the Foundation Axiom. The other point is that

$$(A, M) = (A, N) \iff M = N.$$

The two points combined are used to see that g really is an injection. Finally, define

$$B = g[D]$$

and define $\prec_B$ by

$$g(L) \prec_B g(M) \iff L \prec_D M.$$

Clearly,

$$g : (D, \prec_D) \simeq (B, \prec_B)$$

and

$$g \upharpoonright C : (C, \prec_C) \simeq (A, \prec_A),$$

which completes the proof of Theorem 6.10.   □

To describe the relationship between $(A, \prec_A)$ and $(B, \prec_B)$ in Theorem 6.10, we say that $(B, \prec_B)$ is a *Dedekind completion* of $(A, \prec_A)$. We may define $\mathbb{R}$ to be a Dedekind completion of $\mathbb{Q}$. By the *moreover* part of Theorem 6.10, it is not particularly important which Dedekind completion we choose.

Throughout this chapter, we have been thinking about $\mathbb{Q}$ and $\mathbb{R}$ as certain kinds of linear orderings. But there is much more to numbers than how they are ordered! The rational numbers also come equipped with an arithmetic structure, by which we mean addition, subtraction, multiplication, division, exponentiation, etc. Moreover, the arithmetic structure of the rational numbers lifts nicely to the familiar arithmetic structure for the real numbers. It is possible to explain how this lifting is achieved in terms of left-cuts and Dedekind completions. We already saw a hint of this in our discussion of $\sqrt{2}$ earlier in the chapter. Curious readers might enjoy working out the formal development as an independent project.

Exercises

Exercise 6.1 The point of this exercise is to define $\mathbb{Q}$. Therefore, you may not assume anything about $\mathbb{Q}$, not even that $\mathbb{Q}$ exists, in your solution. However, you may use standard properties of ω with its order, addition and multiplication. For example, writing $\frac{3}{2} = \frac{6}{4}$ is unacceptable at this point because we have not yet defined fractions but writing $3 \cdot 4 = 12 = 2 \cdot 6$ is acceptable because it only refers to natural numbers and multiplication.

Let
$$S = \omega \times (\omega - 1).$$

Here $\omega - 1 = \omega - \{0\} = \{1, 2, 3, \dots\}$. Define a relation E on S by

$$(a, b)\, E\, (c, d) \iff ad = bc.$$

1. Prove that E is an equivalence relation on S.
2. Write $[(a, b)]_E$ for the E-equivalence class of (a, b). That is,

$$[(a, b)]_E = \{(a', b') \mid (a', b')E(a, b)\}.$$

Let $B = S/E$. That is

$$B = \{[(a, b)]_E \mid (a, b) \in S\}.$$

Explain why the formula

$$[(a, b)]_E \prec_B [(c, d)]_E \iff ad < bc$$

defines a relation $\prec_B$ on B. *Hint*: If you think there is nothing to check, think again.

3. Prove that $(B, \prec_B)$ is a countable dense linear ordering with no right endpoint and whose left endpoint is $[(0, 1)]_E$.
4. Let

$$A = B - \{[(0, 1)]_E\}$$

and $\prec_A$ be the restriction of $\prec_B$ to A. Prove that $(A, \prec_A)$ is a countable dense linear ordering without endpoints.

5. Prove that $(A, \succ_A)$ is a countable dense linear ordering without endpoints. (Notice the order is reversed.)
6. Define

$$(\mathbb{Q}, <_{\mathbb{Q}}) = (A, \succ_A)^\frown (B, \prec_B).$$

Prove that $(\mathbb{Q}, <_{\mathbb{Q}})$ is a countable dense linear ordering without endpoints.

Exercise 6.2 Let

$$\mathbb{Z} = \{\ldots, -2, -1, 0, 1, 2, \ldots\}$$

have the usual order on the integers. Prove that $\mathbb{Z} \not\simeq \omega$.

Exercise 6.3 Find a family $\mathcal{F}$ such that

- every element of $\mathcal{F}$ is a countable dense linear ordering, and
- for every countable dense linear ordering $(A, \prec_A)$, there exists a unique $(B, \prec_B) \in \mathcal{F}$ such that $(A, \prec_A) \simeq (B, \prec_B)$.

Exercise 6.4 Complete the proof of the *moreover* part of Theorem 6.10. See the hint given there.

Exercise 6.5 Prove that $\mathbb{R}^\frown \mathbb{R} \not\simeq \mathbb{R}$.

Exercise 6.6   If

$$\mathbb{B} = (B, \vee_\mathbb{B}, \wedge_\mathbb{B}, \neg_\mathbb{B}, \perp_\mathbb{B}, \top_\mathbb{B})$$

is a Boolean algebra, then we call $\mathbb{B}$ *countable* iff B is countably infinite. Prove that if $\mathbb{A}$ and $\mathbb{B}$ are countable Boolean algebras with no atoms, then $\mathbb{A}$ and $\mathbb{B}$ are isomorphic. *Hint*: The solution is quite involved and breaks up into two main components:

- Use a back-and-forth style argument to build an order isomorphism

$$f : (A, \preccurlyeq_\mathbb{A}) \simeq (B, \preccurlyeq_\mathbb{B}).$$

- Prove that every order isomorphism between Boolean algebras is also a Boolean algebra isomorphism. Hence,

$$f : \mathbb{A} \simeq \mathbb{B}.$$

This second part is a general fact about Boolean algebras, so your proof should not use the countable and atomless assumptions.

Here are a couple of useful lemmas you should prove for the back-and-forth part of the argument:

- If f is a finite partial Boolean algebra isomorphism from $\mathbb{A}$ to $\mathbb{B}$, then there exists a finite partial Boolean algebra isomorphism g from $\mathbb{A}$ to $\mathbb{B}$ such that the domain and range of g are finite Boolean algebras. To see this, take $\mathrm{dom}(g)$ to be the Boolean subalgebra of $\mathbb{A}$ generated by $\mathrm{dom}(f)$, and $\mathrm{ran}(g)$ to be the Boolean subalgebra of $\mathbb{B}$ generated by $\mathrm{ran}(f)$. Then extend f in the obvious way to define g.
- The ordering of an atomless Boolean algebra is dense in the sense that if $x \prec z$, then there exists y such that $x \prec y \prec z$. Keep in mind that the ordering is not linear!

We remark that two examples of a countable atomless Boolean algebras were given in Exercises 4.17 and 5.22. By this exercise, they are isomorphic.

7
Filters and ideals

Filters and ideals come up in just about every area of modern mathematics. After some preliminaries, in Section 7.1, we prove Tarski's ultrafilter existence theorem and touch on the theory of ultraproducts. Filters and ideals are particularly important in advanced set theory, where the filter generated by the closed unbounded subsets of an uncountable regular cardinal plays a major role. We give the reader a taste of this sort of infinitary combinatorics in Section 7.2. The main results there are Fodor's theorem and an interesting special case of Solovay's splitting theorem.

7.1 Motivation and definitions

There are many mathematical contexts in which we are given a set X and we talk about *large subsets of X* and *small subsets of X*. This is so common that it is worth writing down what these situations share.

Definition 7.1 Let X be a non-empty set and $\mathcal{F} \subseteq \mathcal{P}(X)$. Then $\mathcal{F}$ is a *filter over X* iff the following conditions hold:

- $\emptyset \notin \mathcal{F}$.
- $X \in \mathcal{F}$.
- For all $A, B \subseteq X$, if $A \in \mathcal{F}$ and $A \subseteq B$, then $B \in \mathcal{F}$.
- For all $A, B \subseteq X$, if $A, B \in \mathcal{F}$, then $A \cap B \in \mathcal{F}$.

To understand the motivation for filters, one can paraphrase the defining conditions as follows.

- *The empty set is not large.*
- *X is large.*
- *If A is large and $A \subseteq B$, then B is also large.*
- *If A and B are large, then so is $A \cap B$.*

If you are not convinced by the last clause, replace *large* by *almost everything* to make it even more believable.

Ideals are to *small* sets what filters are to *large* sets.

Definition 7.2   Let X be a non-empty set and $\mathcal{I} \subseteq \mathcal{P}(X)$. Then $\mathcal{I}$ is an *ideal over X* iff the following conditions hold:

- $\emptyset \in \mathcal{I}$.
- $X \notin \mathcal{I}$.
- For all $A, B \subseteq X$, if $B \in \mathcal{I}$ and $A \subseteq B$, then $A \in \mathcal{I}$.
- For all $A, B \subseteq X$, if $A, B \in \mathcal{I}$, then $A \cup B \in \mathcal{I}$.

In reading the list above, where you see a set belongs to $\mathcal{I}$, you can say out loud that it is *small* (or, maybe better, *almost nothing*) to understand the motivation for the condition.

The next two results explain how filters and ideals are related.

Lemma 7.3   *If $\mathcal{F}$ is a filter over X, then*

$$\{X - A \mid A \in \mathcal{F}\}$$

is an ideal over X.

Lemma 7.4   *If $\mathcal{I}$ is an ideal over X, then*

$$\{X - A \mid A \in \mathcal{I}\}$$

is a filter over X.

So the operation $A \mapsto X - A$ takes filters over X to ideals over X and vice-versa as described by the lemmas. This is an example of what we call *duality* in mathematics.

Is every subset A of X either large or small? It depends on the situation. This thought leads to ultrafilters and prime ideals.

Definition 7.5   Let $\mathcal{F}$ be a filter over X. Then $\mathcal{F}$ is an *ultrafilter over X* iff for every $A \subseteq X$, either $A \in \mathcal{F}$ or $X - A \in \mathcal{F}$.

Definition 7.6   Let $\mathcal{I}$ be an ideal over X. Then $\mathcal{I}$ is a *prime ideal* over X iff for every $A \subseteq X$, either $A \in \mathcal{I}$ or $X - A \in \mathcal{I}$.

It is about time we introduced some examples!

Example If $p \in X$, then $\{A \subseteq X \mid p \in A\}$ is a *principal* ultrafilter over X. This is the least interesting kind of ultrafilter.

Example $\{A \subseteq \omega \mid \omega - A \text{ is finite}\}$ is the *Fréchet* filter over ω. It is not an ultrafilter over ω. For example, neither

$$\text{Even} = \{2n \mid n < \omega\}$$

nor

$$\text{Odd} = \{2n + 1 \mid n < \omega\}$$

are members of the Fréchet filter over ω.

Example For $A \subseteq \omega$, define

$$\text{density}(A) = \lim_{n \to \infty} \frac{|A \cap n|}{n}$$

if the limit exists. Then

$$\{A \subseteq \omega \mid \text{density}(A) = 0\}$$

is the *density* ideal over ω.

Example Let $I = \{x \in \mathbb{R} \mid 0 \leq x \leq 1\}$. For the reader who knows about Lebesgue measure, we mention that an important topic in analysis and probability is the ideal of null sets,

$$\{A \subseteq I \mid A \text{ has Lebesgue measure } 0\},$$

and its dual filter,

$$\{A \subseteq I \mid A \text{ has Lebesgue measure } 1\}.$$

This is not an ultrafilter because there are subsets of I whose Lebesgue measure is strictly between 0 and 1. For example, the interval

$$\{x \in \mathbb{R} \mid 0 \leq x \leq 1/2\}$$

has Lebesgue measure $1/2$.

Exercise 7.1 Let P and X be non-empty sets with $P \subseteq X$, and

$$\mathcal{F} = \{A \subseteq X \mid P \subseteq A\}.$$

1. Prove that $\mathcal{F}$ is a filter over X.
2. Prove that the following are equivalent:

(a) $\mathcal{F}$ is an ultrafilter over X.

(b) P is a singleton.

(c) $\mathcal{F}$ is a principal ultrafilter over X.

Exercise 7.2 Let $\mathcal{F}$ be the Fréchet filter over ω. Suppose that $\mathcal{G}$ is an ultrafilter over ω such that $\mathcal{G} \supseteq \mathcal{F}$. Prove that $\mathcal{G}$ is not principal.

Exercise 7.3 Let X be a non-empty set and $\mathcal{F}$ be an ultrafilter over X. Prove that, for every $n < \omega$ and sequence $\langle A_i \mid i < n \rangle$ of subsets of X, if

$$\bigcup_{i<n} A_i \in \mathcal{F},$$

then there exists $i < n$ such that $A_i \in \mathcal{F}$.

The first important result on this topic is that every filter extends to an ultrafilter. You should notice that the first sentence of the proof uses the Axiom of Choice to know that $\mathcal{P}(X)$ has a cardinality.

Theorem 7.7 (Tarski) *Let $\mathcal{F}$ be a filter over X. Then there exists an ultrafilter $\mathcal{G}$ over X such that $\mathcal{F} \subseteq \mathcal{G}$.*

Proof Let $\kappa = |\mathcal{P}(X)|$ and $\langle A_\alpha \mid \alpha < \kappa \rangle$ be a sequence such that

$$\mathcal{P}(X) = \{A_\alpha \mid \alpha < \kappa\}.$$

We will define a sequence $\langle \mathcal{G}_\alpha \mid \alpha < \kappa \rangle$ by recursion. After each case in our definition of $\mathcal{G}_\beta$, we will verify that if $\alpha < \beta$, then:

- $\mathcal{G}_\beta$ is a filter,
- $A_\alpha \in \mathcal{G}_\beta$ or $X - A_\alpha \in \mathcal{G}_\beta$, and
- $\mathcal{G}_\alpha \subseteq \mathcal{G}_\beta$.

Base case $\beta = 0$.

Define $\mathcal{G}_0 = \mathcal{F}$.

Successor case $\beta = \alpha + 1$.

We break up this case into three subcases. The first subcase is, for every $B \in \mathcal{G}_\alpha$,

$$B \cap A_\alpha \neq \emptyset.$$

In the first subcase, let

$$\mathcal{G}_{\alpha+1} = \{C \subseteq X \mid \text{there exists } B \in \mathcal{G}_\alpha \text{ such that } B \cap A_\alpha \subseteq C\}.$$

Obvserve that $\mathcal{G}_\alpha \cup \{A_\alpha\} \subseteq \mathcal{G}_{\alpha+1}$ because:

- $\mathcal{G}_\alpha \subseteq \mathcal{G}_{\alpha+1}$ since, if $B \in \mathcal{G}_\alpha$, then $B \cap A_\alpha \subseteq B$, so $B \in \mathcal{G}_{\alpha+1}$, and
- $A_\alpha \in \mathcal{G}_{\alpha+1}$ since $X \in \mathcal{G}_\alpha$, so $X \cap A_\alpha \in \mathcal{G}_{\alpha+1}$.

Observe that $\mathcal{G}_{\alpha+1}$ is a filter because:

- $\emptyset \notin \mathcal{G}_{\alpha+1}$ by the first subcase hypothesis,
- $X \in \mathcal{G}_\alpha$ since $X \cap A_\alpha \subseteq X$,
- $\mathcal{G}_{\alpha+1}$ is clearly closed upward under $\subseteq$, and
- $\mathcal{G}_{\alpha+1}$ is closed under pairwise intersections since if $B, C \in \mathcal{G}_\alpha$, then

$$(B \cap A_\alpha) \cap (C \cap A_\alpha) = (B \cap C) \cap A_\alpha \in \mathcal{G}_{\alpha+1}.$$

The second subcase is that the first subcase fails and, for every $B \in \mathcal{G}_\alpha$,

$$B \cap (X - A_\alpha) \neq \emptyset.$$

In the second subcase, let

$$\mathcal{G}_{\alpha+1} = \{C \subseteq X \mid \text{there exists } B \in \mathcal{G}_\alpha \text{ such that } B \cap (X - A_\alpha) \subseteq C\}.$$

Much like in the first subcase, one shows that

$$\mathcal{G}_\alpha \cup \{X - A_\alpha\} \subseteq \mathcal{G}_{\alpha+1}$$

and $\mathcal{G}_{\alpha+1}$ is a filter. The reader should complete the verification.

The third subcase is that the first and second subcases fail. We will show this does not happen. For contradiction, suppose

$$B, C \in \mathcal{G}_\alpha,$$

$$B \cap A_\alpha = \emptyset$$

and

$$C \cap (X - A_\alpha) = \emptyset.$$

Let $D = B \cap C$. Then

$$D \in \mathcal{G}_\alpha,$$

$$D \cap A_\alpha = \emptyset$$

and

$$D \cap (X - A_\alpha) = \emptyset.$$

Therefore $D = \emptyset$. But $\emptyset \notin \mathcal{G}_\alpha$ since $\mathcal{G}_\alpha$ is a filter.

Limit case β *is a limit ordinal.*

Define

$$\mathcal{G}_\beta = \bigcup_{\alpha < \beta} \mathcal{G}_\alpha.$$

It is easy to check that $\mathcal{G}_\beta$ is a filter because it is the union of a $\subseteq$-increasing sequence of filters. Also, if $\alpha < \beta$, then either

$$A_\alpha \in \mathcal{G}_{\alpha+1} \subseteq \mathcal{G}_\beta$$

or

$$(X - A_\alpha) \in \mathcal{G}_{\alpha+1} \subseteq \mathcal{G}_\beta.$$

That completes the recursive definition of the sequence

$$\langle \mathcal{G}_\alpha \mid \alpha < \kappa \rangle$$

and the verification that it has the desired properties. Now let

$$\mathcal{G} = \bigcup_{\alpha < \kappa} \mathcal{G}_\alpha.$$

As in the limit case, we see that $\mathcal{G}$ is a filter. Clearly, $\mathcal{F} = \mathcal{G}_0 \subseteq \mathcal{G}$. Finally, $\mathcal{G}$ is an ultrafilter because, for every $\alpha < \kappa$, either

$$A_\alpha \in \mathcal{G}_{\alpha+1} \subseteq \mathcal{G}$$

or

$$(X - A_\alpha) \in \mathcal{G}_{\alpha+1} \subseteq \mathcal{G}.$$

$\square$

Exercise 7.4 Consider the special case of Theorem 7.7 in which $X = \omega$ and $\mathcal{F}$ is the Fréchet filter over ω. Notice that in the proof

$$\kappa = |\mathcal{P}(\omega)| = 2^{\aleph_0}.$$

Let $\langle \mathcal{G}_\alpha \mid \alpha < 2^{\aleph_0} \rangle$ be the sequence of filters extending $\mathcal{F}$ that was recursively constructed in the proof of Theorem 7.7.

1. Prove by induction on $\alpha < 2^{\aleph_0}$ that $|\mathcal{G}_\alpha| < 2^{\aleph_0}$ and $\mathcal{G}_\alpha$ is not an ultrafilter over ω.

2. Use ideas similar to the proof of Theorem 7.7 to show that there are $2^{2^{\aleph_0}}$ many non-principal ultrafilters over ω. *Hint*: This is a slightly challenging exercise. View $\langle \mathcal{G}_\alpha \mid \alpha < 2^{\aleph_0} \rangle$ as a branch through a certain kind of tree with $2^{\aleph_0}$ many levels. Argue that the tree has $2^{2^{\aleph_0}}$ many distinct branches each of which corresponds to a different ultrafilter over ω.

Exercise 7.5 below introduces the reader to a certain construction, which is known as *taking an ultrapower by an ultrafilter*. Intuitively, the idea is to start with a sequence of structures (in the exercise, the structures are linear orderings) and an ultrafilter, $\mathcal{F}$, and to form a new structure by averaging out according to $\mathcal{F}$. Our meaning will become clear when the reader does the exercise and reads the discussion that follows. Before starting, recall how products of sets were defined in Exercise 4.10. In particular, given a sequence $\langle A_n \mid n < \omega \rangle$, we define $\prod_{n<\omega} A_n$ to be the set of functions f such that $\operatorname{dom}(f) = \omega$ and, for every $n < \omega$,

$$f(n) \in A_n.$$

In the special case where all the A_ns are the same, say $A_n = B$, we end up with

$$\prod_{n<\omega} A_n = \prod_{n<\omega} B = \{f \mid f \text{ is a function from } \omega \text{ to } B\} = {}^\omega B.$$

Immediately after Exercise 7.5, there is a long discussion of the significance of the ultrapower construction.

Exercise 7.5   Let $\langle A_n \mid n < \omega \rangle$ be a sequence and

$$P = \prod_{n<\omega} A_n.$$

Let $\mathcal{F}$ be an ultrafilter over ω.

1. Define a relation $\sim$ on P by

$$f \sim g \iff \{n < \omega \mid f(n) = g(n)\} \in \mathcal{F}.$$

   Prove that $\sim$ is an equivalence relation on P.
2. For $f \in P$, let

$$[f] = \{g \in P \mid f \sim g\}.$$

   Also, let

$$A = \{[f] \mid f \in P\}.$$

Assume that, for each $n < \omega$, we are given a relation

$$R_n \subseteq A_n \times A_n.$$

Prove that we may define a relation

$$R \subseteq A \times A$$

by setting

$$[f] \ R \ [g] \iff \{n < \omega \mid f(n) \ R_n \ g(n)\} \in \mathcal{F}.$$

In other words, prove that the definition of R does not depend on the choice of representatives for the equivalence classes $[f]$ and $[g]$.

3. Assume that (A_n, R_n) is a strict linear ordering for every $n < \omega$. Prove that (A, R) is also a strict linear ordering.

4. Now assume that, for every $n < \omega$,

$$(A_n, R_n) = (\omega, <)$$

where $<$ is the usual order on the natural numbers. Suppose that $\mathcal{F}$ is a non-principal ultrafilter over ω. Prove that (A, R) is not a wellordering. *Hint*: Notice that, in this case, $P = {}^\omega\omega$. For $c \in \mathbb{Z}$, consider the function

$$f_c : \omega \to \omega$$

defined by

$$f_c(n) = \begin{cases} n + c & \text{if } n + c \geq 0 \\ 0 & \text{otherwise.} \end{cases}$$

Prove that

$$\cdots \ R \ [f_{-3}] \ R \ [f_{-2}] \ R \ [f_{-1}] \ R \ [f_0].$$

5. Again assume that, for every $n < \omega$,

$$(A_n, R_n) = (\omega, <).$$

But suppose instead that $\mathcal{F}$ is a principal ultrafilter over ω. Prove that

$$(A, R) \simeq (\omega, <).$$

Hint: Most but not all of the details are contained in the discussion after this exercise, so read it first if you get stuck.

That completes the instructions for Exercise 7.5 but there is much more we should tell the reader about the construction done there. The pair (A, R) is called the *ultraproduct* of the sequence $\langle (A_n, R_n) \mid n < \omega \rangle$ by $\mathcal{F}$. A popular way to express the definition of R is

$$[f] \; R \; [g] \iff f(n) \; R_n \; g(n) \text{ for } \mathcal{F}\text{-almost every } n < \omega.$$

Often, it helps to think about ultraproducts using this alternative language. This language makes it clearer what we meant by *averaging out* in the paragraph preceding Exericse 7.5. In this exercise, we took ultraproducts of linear orderings but it is also possible to take ultraproducts of other kinds of structures. For example, the reader should see how the ultraproduct of a sequence of Boolean algebras would be defined.

The example of an ultraproduct (A, R) given in the last two parts of Exercise 7.5 is called the *ultrapower* of $(\omega, <)$ by $\mathcal{F}$. Instead of saying *ultrapower*, we use the term *ultraproduct* in this case because all of the pairs (A_n, R_n) are the same. The main point of part 4 of Exercise 7.5 is that the ultraproduct of wellorderings need not be a wellordering. Figure 7.1 is a rough picture of what (A, R) looks like in this case. The initial segment of (A, R) that is isomorphic to ω is really:

$$[n \mapsto 0] \; R \; [n \mapsto 1] \; R \; [n \mapsto 2] \; R \; \cdots$$

Keep in mind that, for a fixed $c < \omega$, the function $n \mapsto c$ is the function with the constant value c. The chain of relations that we just displayed says that, for every constant $c < \omega$,

$$[n \mapsto c] \; R \; [n \mapsto c + 1].$$

This is because

$$\{n < \omega \mid c < c + 1\} = \omega \in \mathcal{F}.$$

But we are also claiming that there is no equivalence class strictly between each $[n \mapsto c]$ and $[n \mapsto c + 1]$. To see this, let $b = c + 1$ and suppose $f : \omega \to \omega$ is a function such that

$$[f] \; R \; [n \mapsto b].$$

Then

$$\{n < \omega \mid f(n) < b\} \in \mathcal{F}.$$

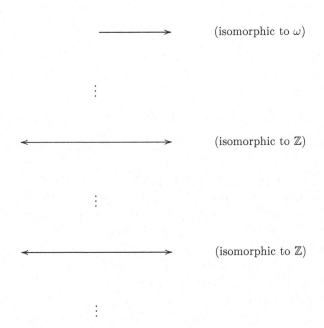

Figure 7.1 An ultrapower of $(\omega, <)$ by a non-principal ultrafilter

Clearly,

$$\{n < \omega \mid f(n) < b\} = \bigcup_{a<b}\{n < \omega \mid f(n) = a\}.$$

By Exercise 7.3, there exists $a < b$ such that

$$\{n < \omega \mid f(n) = a\} \in \mathcal{F}.$$

By definition, this just says that

$$[f] = [n \mapsto a].$$

Finally, observe that $a \leq c$.

Looking again at Figure 7.1, after the initial segment of (A, R) that is isomorphic to ω, there are infinitely many pieces each of which is isomorphic to $\mathbb{Z}$. We call these $\mathbb{Z}$-*chains*. In terms of the functions f_c defined in part 4 of Exercise 7.5, here is one example of a Z-chain:

$$\cdots R \; [f_{-2}] \; R \; [f_{-1}] \; R \; [f_0] \; R \; [f_1] \; R \; [f_2] \; R \; \cdots .$$

Using the ideas from the previous paragraph, the reader should prove that $[f_c] \; R \; [f_{c+1}]$ and there are no equivalence classes of functions strictly between $[f_c]$ and $[f_{c+1}]$. In other words, that this really is an example of a Z-chain.

Now define

$$h_c(n) = \begin{cases} 3n + c & \text{if } 3n + c \geq 0 \\ 0 & \text{otherwise.} \end{cases}$$

Here is a second example of a Z-chain:

$$\cdots R \; [h_{-2}] \; R \; [h_{-1}] \; R \; [h_0] \; R \; [h_1] \; R \; [h_2] \; R \; \cdots$$

The reader should verify that this is indeed a Z-chain. We claim that this Z-chain lies entirely after our first example of a Z-chain. In other words, for all $a, c \in \mathbb{Z}$,

$$[f_a] \; R \; [h_c].$$

By the definition of R, this just says that

$$\{n < \omega \mid f_a(n) < h_c(n)\} \in \mathcal{F}.$$

The main observation needed to see why this is true is that

$$\{n < \omega \mid n + a < 3n + c\} = \{n < \omega \mid (a - c)/2 < n\}$$
$$= \omega - \{n < \omega \mid n \leq (a - c)/2\}$$
$$\in \mathcal{F}$$

because it is the complement of a finite set and $\mathcal{F}$ is non-principal.

We give a third example of a Z-chain. For $b \in \mathbb{Z}$, define

$$g_b(n) = \begin{cases} 2n + b & \text{if } 2n + b \geq 0 \\ 0 & \text{otherwise.} \end{cases}$$

Our third Z-chain

$$\cdots R \; [g_{-2}] \; R \; [g_{-1}] \; R \; [g_0] \; R \; [g_1] \; R \; [g_2] \; R \; \cdots$$

lies strictly between the other two as the reader should verify.

Building on these observations, one sees that the Z-chains themselves form a dense linear ordering without endpoints. We leave it to the reader to fill in the details and continue this investigation as an extremely worthwhile project.

Ultraproducts are used in many branches of mathematics, not just set theory. A famous and intriguing example is Abraham Robinson's theory of non-standard analysis, which rehabilitated Gottfried Wilhelm Leibniz's seventeenth-century infinitesimal calculus. *Infinitesimals* are supposed to be numbers $\varepsilon > 0$ such that $\varepsilon < x$ for every positive number $x \in \mathbb{R}$. Of course, there are no such $\varepsilon \in \mathbb{R}$. Nevertheless, without really knowing what he meant by infinitesimals, Leibniz developed recipes for working with them that yielded correct answers to questions about geometry and physics. While this represented tremendous intuition, Leibniz's theory was considered controversial and was eventually abandoned in favor of the rigorous development of calculus provided by Augustin-Louis Cauchy in the eighteenth-century. Much later, in the 1960s, Robinson vindicated Leibniz by saying what infinitesimals really are and explaining why it was legitimate to derive calculus formulas using them. To get the idea, let $\mathcal{F}$ be a non-principal ultrafilter over ω and take the ultrapower of $(\mathbb{R}, <)$ by $\mathcal{F}$. Call this ultrapower $(\mathbb{R}^*, <^*)$. Pretty much like in our elaboration on Exercise 7.5, we see that, for all constants $a < b$ in $\mathbb{R}$,

$$[n \mapsto a] <^* [n \mapsto b].$$

So there is a copy of $(\mathbb{R}, <)$ sitting inside of $(\mathbb{R}^*, <^*)$. But there are new points to the left, to the right and in between. For example, for every $c \in \mathbb{R}$,

$$[n \mapsto c] <^* [n \mapsto n],$$

which shows there are new *positive infinite* members of $\mathbb{R}^*$ entirely to the right of our copy of $\mathbb{R}$. Similarly, for every $c \in \mathbb{R}$,

$$[n \mapsto -n] <^* [n \mapsto c],$$

so there are new *negative infinite* members of $\mathbb{R}^*$ entirely to the left of our copy of $\mathbb{R}$. Even more interesting is the fact that, for every positive $c \in \mathbb{R}$,

$$[n \mapsto 0] <^* [n \mapsto 1/n] <^* [n \mapsto c].$$

In other words, $[n \mapsto 1/n]$ is greater than our copy of 0 and less than our copy of c for every positive real number c. For this reason, it is reasonable to say that the equivalence class $[n \mapsto 1/n]$ is an example of an *infinitesimal* member of $\mathbb{R}^*$. (Technically, when we write $n \mapsto 1/n$ here, we really mean

$$n \mapsto \begin{cases} 1/n & \text{if } n \neq 0 \\ 0 & \text{if } n = 0 \end{cases}$$

because the domain must be ω but we cannot divide by zero.) There is more to understanding why one can reason about $(\mathbb{R}^*, <^*)$ and come to certain correct conclusions about $(\mathbb{R}, <)$ but that requires a basic background in mathematical logic, which we do not presume. Our intent was merely to introduce the reader to this fascinating and historically significant subject.

7.2 Club and stationary sets

This section builds on the previous section and Chapters 3 and 4. Given a limit ordinal κ, there is a interesting and very useful filter over κ called the club filter. Its dual is called the non-stationary ideal. Before we can say what these are, we need some definitions.

Definition 7.8 Let κ be a limit ordinal and C be a set. Then:

• C is *unbounded in* κ iff $\sup(C \cap \kappa) = \kappa$.
• C is *closed in* κ iff for every $\alpha < \kappa$, if $C \cap \alpha \neq \emptyset$, then

$$\sup(C \cap \alpha) \in C.$$

• C is *club in* κ iff C is closed and unbounded in κ.

Here are some examples with $\kappa = \omega_1$.

Example If $\alpha < \omega_1$, then $\{\beta < \kappa \mid \alpha < \beta\}$ is club in ω_1.

Example $\{\alpha < \omega_1 \mid \alpha$ is a limit ordinal$\}$ is club in ω_1. It is closed because a limit of limit ordinals is also a limit ordinal. It is unbounded because if α is a countable ordinal, then $\alpha + \omega$ is a countable limit ordinal.

Example $\{\alpha < \omega_1 \mid \alpha$ is a successor ordinal$\}$ is not club in ω_1. While it is unbounded, it is not closed. For example,

$$\omega = \sup_{n < \omega}(n + 1)$$

but ω is not a successor ordinal.

Example $\{\alpha < \omega_1 \mid \omega^\alpha = \alpha\}$ is club in ω_1. Here ω^α is ordinal exponentiation. The proof is a little more than what you were asked to show in Exercise 4.12.

Definition 7.8 applies to all limit ordinals but most often we apply it specifically to uncountable regular cardinals such as ω_1. Remember that κ is regular iff $\mathrm{cf}(\kappa) = \kappa$. And remember that all infinite successor cardinals are regular.

Exercise 7.6 Let κ be a regular cardinal and $C \subseteq \kappa$. Prove C is unbounded in κ iff $|C| = \kappa$.

Now we say what this has to do with filters.

Lemma 7.9 *Assume that* $\mathrm{cf}(\kappa) > \omega$. *Let*

$$\mathcal{F} = \{A \subseteq \kappa \mid \text{there exists a club } C \text{ in } \kappa \text{ such that } C \subseteq A\}.$$

Then $\mathcal{F}$ *is a filter over* κ.

Proof The only condition in the definition of filter that is not obvious is closure under intersections. It is enough to show that if C and D are club in κ, then so is $C \cap D$.

First we show that $C \cap D$ is closed in κ. Let $\beta < \kappa$ and assume that $C \cap D \cap \beta \neq \emptyset$. Let

$$\alpha = \sup(C \cap D \cap \beta).$$

We must show that $\alpha \in C \cap D$. Easily, we see that $\alpha = \sup(C \cap \alpha)$ and $\alpha = \sup(D \cap \alpha)$. Since C and D are closed, $\alpha \in C$ and $\alpha \in D$.

To finish we show that $C \cap D$ is unbounded in κ. Let $\alpha < \kappa$. We must find $\delta \in C \cap D$ such that $\alpha < \delta$. By recursion on $n < \omega$, define ordinals $\beta_n, \gamma_n < \kappa$ as follows. Pick $\beta_0 \in C$ with $\beta_0 > \alpha$. Given β_n, pick $\gamma_n \in D$ with $\gamma_n > \beta_n$. Given γ_n, pick $\beta_{n+1} \in C$ with $\beta_{n+1} > \gamma_n$. We can do all this picking because C and D are unbounded in κ. Now let δ be the supremum of either sequence; it is the same because of the interleaving. That is,

$$\delta = \sup_{n<\omega} \beta_n = \sup_{n<\omega} \gamma_n.$$

Then $\delta < \kappa$ because $\mathrm{cf}(\kappa) > \omega$. Since C and D are closed,

$$\delta = \sup(C \cap \delta) \in C$$

and

$$\delta = \sup(D \cap \delta) \in D,$$

hence $\delta \in C \cap D$ as desired. □

The filter in Lemma 7.9 is called the *club filter over κ*. The main point of the proof is that the intersection of two clubs is club if κ has uncountable cofinality. This is not necessarily true if κ has countable cofinality. For instance, Even $= \{2n \mid n < \omega\}$ and Odd $= \{2n + 1 \mid n < \omega\}$ are disjoint unbounded subsets of ω, and these sets are closed for trivial reasons.

Exercise 7.7 Let κ be an uncountable regular cardinal and

$$f : \kappa \to \kappa$$

be a function. Prove that

$$\{\alpha < \kappa \mid f[\alpha] \subseteq \alpha\}$$

is club in κ. *Remark:* This implies that $\{\alpha < \omega_1 \mid \omega^\alpha = \alpha\}$ is club in ω_1 in the special case $\kappa = \omega_1$ and $f : \alpha \mapsto \omega^\alpha$. The special case was the subject of Exercise 4.12 and the general argument is similar.

Exercise 7.8 Let κ be an uncountable regular cardinal. Prove that if $\theta < \kappa$ and $\langle C_\alpha \mid \alpha < \theta \rangle$ is a sequence of club subsets of κ, then the set

$$\bigcap \{C_\alpha \mid \alpha < \theta\}$$

is club in κ. *Hint:* Use induction on $\theta < \kappa$. The successor case, $\theta = \eta + 1$, is immediate from the induction hypothesis and the case $\theta = 2$, which was handled in the proof of Lemma 7.9. Suppose θ is a limit ordinal. In this case, the proof of closure is straightforward (similar to the case $\theta = 2$). For the proof of unboundedness, given $\beta_0 < \kappa$, define an increasing sequence $\langle \beta_\eta \mid \eta < \theta \rangle$ such that, for every $\eta < \theta$,

$$\beta_\eta \in \bigcap \{C_\zeta \mid \zeta < \eta\}.$$

Exercise 7.9 Let κ be a regular cardinal. Give an example of a sequence $\langle C_\alpha \mid \alpha < \kappa \rangle$ such that, for every $\alpha < \kappa$, C_α is club in κ but

$$\bigcap \{C_\alpha \mid \alpha < \kappa\} = \emptyset.$$

Let $\mathcal{F}$ be the club filter over κ. Recall that

$$\mathcal{F} = \{S \subseteq \kappa \mid \text{there exists a club } C \text{ in } \kappa \text{ such that } C \subseteq S\}.$$

As we indicated before, we intuitively think of members of $\mathcal{F}$ as large subsets of κ. Let $\mathcal{I}$ be the ideal dual to $\mathcal{F}$. Then

$$\mathcal{I} = \{S \subseteq \kappa \mid \kappa - S \in \mathcal{F}\}.$$

We think of members of $\mathcal{I}$ as small subsets of κ. Observe that if $S \subseteq \kappa$, then

$$S \in \mathcal{I} \iff \text{there is a club } C \text{ in } \kappa \text{ such that } S \cap C = \emptyset.$$

We are also interested in subsets of κ that are not small. Notice that if $S \subseteq \kappa$, then

$$S \notin \mathcal{I} \iff \text{for every club } C \text{ in } \kappa, \ S \cap C \neq \emptyset.$$

Intuitively, this says that a subset of κ is not small iff it meets every large subset of κ. We give such sets the following name.

Definition 7.10 Let κ be a limit ordinal and $S \subseteq \kappa$. Then S is *stationary* in κ iff for every club C in κ,

$$S \cap C \neq \emptyset.$$

Notice that if C is club, then C is stationary. This is because if D is club, then $C \cap D$ is club, in particular, $C \cap D \neq \emptyset$. Intuitively, this says that if a set is large, then it is not small.

You can get additional intuition for Definition 7.10 if you happen to know about Lebesgue measure on the unit interval

$$\{x \in \mathbb{R} \mid 0 \leq x \leq 1\}.$$

The relevant analogies are:

$$\frac{\text{contains a club}}{\text{measure 1}} = \frac{\text{stationary}}{\text{positive measure}} = \frac{\text{not stationary}}{\text{measure 0}}$$

The following exercise gives an important example with $\kappa = \omega_2$.

Exercise 7.10 Let

$$C = \{\alpha < \omega_2 \mid \alpha \text{ is a limit ordinal}\},$$

$$E = \{\alpha \in C \mid \text{cf}(\alpha) = \omega\}$$

and

$$C - E = \{\alpha \in C \mid \text{cf}(\alpha) = \omega_1\}.$$

1. Prove that C is club in ω_2. (This is pretty obvious.)
2. Prove that E is stationary in ω_2.
3. Prove that $C - E$ is stationary in ω_2.
4. Use parts 1, 2 and 3 to prove that the club filter over ω_2 is not an ultrafilter.

Now we come to one of the most fundamental tools for studying club and stationary sets.

Theorem 7.11 (Fodor)   *Let κ be uncountable regular cardinal and $f : \kappa \to \kappa$ be a function such that*

$$\{\alpha < \kappa \mid f(\alpha) < \alpha\}$$

is stationary in κ. Then there exists $\theta < \kappa$ such that

$$\{\alpha < \kappa \mid f(\alpha) = \theta\}$$

is stationary in κ.

We will derive Theorem 7.11 from Lemma 7.12, which is interesting in its own right. Exercise 7.9 tells us that the intersection of κ many club sets might not be club. Lemma 7.12 says that the *diagonal intersection* of κ many club sets is club.

Lemma 7.12   *Let κ be an uncountable regular cardinal. Suppose that*

$$\langle C_\alpha \mid \alpha < \kappa \rangle$$

is a sequence of club subsets of κ. Let

$$D = \{\alpha < \kappa \mid \alpha \in C_\theta \text{ for every } \theta < \alpha\}.$$

Then D is club in κ.

We call D the *diagonal intersection* of $\langle C_\alpha \mid \alpha < \kappa \rangle$. Most commonly, you will see it written $\Delta_{\alpha<\kappa} C_\alpha$.

Proof of Lemma 7.12   First we show that D is closed in κ. Let $\gamma < \kappa$. Assume that $D \cap \gamma \neq \emptyset$ and let $\beta = \sup(D \cap \gamma)$. We must show that $\beta \in D$. For contradiction, suppose that $\beta \notin D$. It follows easily that β is a limit ordinal and $D \cap \beta = D \cap \gamma$. Hence

$$\beta = \sup(D \cap \beta).$$

By the definition of D, there exists $\theta < \beta$ such that $\beta \notin C_\theta$. Since C_θ is closed, there are two cases:

1. $C_\theta \cap \beta = \emptyset$.
2. $C_\theta \cap \beta \neq \emptyset$ and $\sup(C_\theta \cap \beta) \in C_\theta$.

In the second case, $\sup(C_\theta \cap \beta) < \beta$ because $\beta \notin C_\theta$. In either case, we may pick $\alpha \in D$ such that $\theta < \alpha < \beta$. In the second case, we can also make sure that $\sup(C_\theta \cap \beta) < \alpha$. Then, in both cases, $\alpha \in D$ and $\theta < \alpha$ but $\alpha \notin C_\theta$. This directly contradicts the definition of D.

Now we show that D is unbounded in κ. By recursion on $\theta < \kappa$ define β_θ as follows. Let $\beta_0 = 1$. Given $\beta_\theta < \kappa$, pick $\beta_{\theta+1} > \beta_\theta$ such that

$$\beta_{\theta+1} \in \bigcap_{\eta < \beta_\theta} C_\eta.$$

This is possible by Exercise 7.8. If θ is a limit ordinal, then let

$$\beta_\theta = \sup_{\eta < \theta} \beta_\eta.$$

That completes the recursive definition of $\langle \beta_\theta \mid \theta < \kappa \rangle$. By induction on $\theta < \kappa$, one sees that, for every $\eta < \theta$,

$$\beta_\eta < \beta_\theta$$

and if $\zeta < \beta_\eta$, then

$$\beta_\theta \in C_\zeta.$$

Suppose that $\theta < \kappa$ is a limit ordinal. Then, for every $\zeta < \beta_\theta$,

$$\{\beta_\eta \mid \zeta + 1 < \eta < \theta\} \subseteq C_\zeta$$

and

$$\beta_\theta = \sup(\{\beta_\eta \mid \zeta + 1 < \eta < \theta\}) = \sup(C_\zeta \cap \beta_\theta)$$

so

$$\beta_\theta \in C_\zeta$$

since C_ζ is closed. We have seen that

$$\{\beta_\theta \mid \theta < \kappa \text{ and } \theta \text{ is a limit ordinal}\} \subseteq D.$$

The set on the left is unbounded in κ and hence so is D. $\square$

Proof of Theorem 7.11 Let

$$S = \{\alpha < \kappa \mid f(\alpha) < \alpha\}.$$

Our assumption is that S is stationary in κ. For each $\theta < \kappa$, let

$$T_\theta = \{\alpha \in S \mid f(\alpha) = \theta\}.$$

For contradiction, suppose that no T_θ is stationary in κ. For each $\theta < \kappa$, pick C_θ club in κ such that $T_\theta \cap C_\theta = \emptyset$. Let

$$D = \{\alpha < \kappa \mid \alpha \in C_\theta \text{ for every } \theta < \alpha\}.$$

By Lemma 7.12, D is club in κ. Pick $\alpha \in D \cap S$. Then, for every $\theta < \alpha$, $f(\alpha) \neq \theta$. In other words, $f(\alpha) \geq \alpha$. But $f(\alpha) < \alpha$ since $\alpha \in S$. □

In Exercise 7.10, we saw that the club filter over ω_2 is not an ultrafilter. The proof outlined there generalizes to regular cardinals $\lambda \geq \omega_2$. To see this, note that, for every regular cardinal $\kappa < \lambda$,

$$\{\alpha < \lambda \mid \mathrm{cf}(\alpha) = \kappa\}$$

is stationary in λ. Moreover, these sets are disjoint for different κ. But Exercise 7.10 does not generalize to $\lambda = \omega_1$ because ω is the only regular cardinal less than ω_1. However, the following theorem implies that the club filter is not an ultrafilter over ω_1. It is a special case of a more powerful result known as the Solovay splitting theorem.

Theorem 7.13   *There is $S \subseteq \omega_1$ such that S and $\omega_1 - S$ are stationary in ω_1.*

Proof   Suppose Theorem 7.13 is false. For each positive $\alpha < \omega_1$, pick a surjection

$$f_\alpha : \omega \to \alpha.$$

This is obviously possible because α is countable. Intuitively, we will reach a contradiction by using the club filter over ω_1 to average out the sequence $\langle f_\alpha \mid \alpha < \omega_1 \rangle$ and obtain a new surjection from ω onto ω_1. This will be a contradiction because ω_1 is uncountable by definition.

Claim 7.13.1   *For every $n < \omega$, there exists $\theta < \omega_1$ such that*

$$\{\alpha < \omega_1 \mid f_\alpha(n) = \theta\}$$

is stationary in ω_1.

Proof Fix $n < \omega$. Consider the function

$$g : \alpha \mapsto f_\alpha(n).$$

Then $g(\alpha) < \alpha$ whenever $0 < \alpha < \omega_1$. By Fodor's theorem, there exists $\theta < \omega_1$ such that

$$\{\alpha < \omega_1 \mid g(\alpha) = \theta\}$$

is stationary in ω_1. □

Claim 7.13.2 *For every $n < \omega$, there is at most one $\theta < \omega_1$ such that*

$$\{\alpha < \omega_1 \mid f_\alpha(n) = \theta\}$$

is stationary in ω_1.

Proof Fix $n < \omega$. For $\theta < \omega_1$, let

$$S_\theta = \{\alpha < \omega_1 \mid f_\alpha(n) = \theta\}.$$

Clearly, if $\eta < \theta < \omega_1$, then

$$S_\eta \cap S_\theta = \emptyset.$$

Suppose $\eta < \theta < \omega_1$ and both S_η and S_θ are stationary. Then $\omega_1 - S_\eta$ is also stationary because

$$S_\theta \subseteq \omega_1 - S_\eta.$$

This means the statement of Theorem 7.13 holds with $S = S_\eta$. But we assumed that Theorem 7.13 is false, so we have a contradiction. □

Claims 7.13.1 and 7.13.2 allow us to define a function $g : \omega \to \omega_1$ by setting $g(n)$ equal to the unique $\theta < \omega_1$ such that

$$\{\alpha < \omega_1 \mid f_\alpha(n) = \theta\}$$

is stationary in ω_1.

Claim 7.13.3 *g is a surjection from ω to ω_1.*

Proof Let $\theta < \omega_1$. For contradiction, suppose that $\theta \notin \text{ran}(g)$. Then, for every $n < \omega$,

$$\{\alpha < \omega_1 \mid f_\alpha(n) = \theta\}$$

is non-stationary in ω_1. For each $n < \omega$, pick a club C_n in ω_1 such that

$$C_n \cap \{\alpha < \omega_1 \mid f_\alpha(n) = \theta\} = \emptyset.$$

In other words, for every $n \in \omega$ and every $\alpha \in C_n$,

$$f_\alpha(n) \neq \theta.$$

Let

$$D = \bigcap \{C_n \mid n < \omega\}.$$

Then D is club in ω_1 and, for every $n < \omega$ and every $\alpha \in D$,

$$f_\alpha(n) \neq \theta.$$

Since D is unbounded in ω_1, there exists $\alpha \in D$ such that $\theta < \alpha$. Since f_α is a surjection from ω to α, there exists $n < \omega$ such that

$$f_\alpha(n) = \theta.$$

This is a contradiction. □

Claim 7.13.3 contradicts the fact that ω_1 is uncountable. □

Exercise 7.11 Use Theorem 7.13 to prove that the club filter over ω_1 is not an ultrafilter.

Solovay's splitting theorem says that if κ is a regular uncountable cardinal and S is a stationary in κ, then there is a sequence $\langle S_\alpha \mid \alpha < \kappa \rangle$ of stationary subsets of S such that, for all $\alpha < \beta < \kappa$,

$$S_\alpha \cap S_\beta = \emptyset.$$

In other words, S can be split into κ many disjoint stationary pieces. This is more powerful than Theorem 7.13, which says that ω_1 can be split into two disjoint stationary pieces.

Appendix
Summary of exercises on Boolean algebra

Boolean algebras were defined just before Exercises 2.12. That exercise gave a characterization of finite Boolean algebras up to isomorphism, namely, they all look like $\mathcal{P}(S)$ for some finite set S. The proof involved looking at atoms.

An example of an infinite atomless Boolean algebra, $\mathcal{P}(\omega)/\text{Finite}$, was given in Exercise 2.13. By Exercise 4.15, $\mathcal{P}(\omega)/\text{Finite}$ is uncountable. Another example of an uncountable atomless Boolean algebra, the family of clopen subsets of the Baire space, was the subject of Exercise 5.23.

The finite Boolean algebras of truth tables, $\mathbb{T}_n$ for $n < \omega$, and the infinite Boolean algebra $\mathbb{T}_\infty$ where discussed in Exercise 4.17. We saw that $\mathbb{T}_\infty$ is a countable atomless Boolean algebra in part 6 of that exercise. Another example of a countable atomless Boolean algebra was given in Exercise 5.22. This was the family of clopen subsets of the Cantor space. The fact that all countable atomless Boolean algebras are isomorphic was the topic of Exercise 6.6. This is an important theorem whose proof uses a back-and-forth construction.

Selected further reading

There are several other undergraduate textbooks on set theory. For example, some of the material that we covered in our course can also be found in Enderton (1977) and Hrbacek and Jech (1999).[1] These books are different from each other and from ours, which certainly benefits the reader.

Those who would like to go on to more advanced set theory should first learn basic mathematical logic. Again, there are many options. To name two, Enderton (2001) is an excellent starting point, while Goldstern and Judah (1998) is a bit more advanced.

This course and an elementary background in logic prepare the reader for graduate level set theory. Two indispensable texts are Kunen (1983) and Jech (2003). Our reader who enjoyed ordinal and cardinal arithmetic and infinitary combinatorics, especially Sections 4.3, 5.5, 5.6 and 7.2, and would like to learn the relative consistency results of Gödel and Cohen on the Continuum Hypothesis, which were mentioned in Section 4.2, will be particularly drawn to these wonderful classics.

The material covered in Sections 5.1 through 5.4 is part of a broad subject called *descriptive set theory*, which is a certain combination of set theory, analysis and logic. To continue in this direction, the reader would want to know the fundamentals of analysis. At the advanced undergraduate level, two analysis textbooks to consider are Rudin (1976) and Royden (1988). For descriptive set theory, two beginning graduate level texts are Kechris (1995) and Moschovakis (2009). These have very different emphases. Roughly,

---

[1] Here and below, we cite only the most recent edition available.

the former ties set theory to analysis more than logic, while for the latter it is the other way around.

Yet another subject that is intertwined with set theory and logic is *model theory*. In Chapter 6, we saw examples of *classification up to isomorphism*. This idea is important throughout mathematics but especially in model theory. We also touched on ultraproducts in Section 7.1. This is a model-theoretic construction that has applications in many fields, particularly in set theory. A classic beginning graduate model theory text is Chang and Keisler (1990).

Set theory is a vast topic of current mathematical research. The enormous *Handbook of set theory*, edited by Foreman and Kanamori (2010), comes in three volumes with a total of twenty-four chapters by various authors. It suffices to give the reader an accurate impression of the many directions the subject has taken in recent decades.

Bibliography

Chang, C. C., and Keisler, H. J. 1990. *Model theory.* Third edn. Studies in Logic and the Foundations of Mathematics, vol. 73. Amsterdam: North-Holland.

Enderton, Herbert B. 1977. *Elements of set theory.* New York: Academic Press [Harcourt Brace Jovanovich Publishers].

Enderton, Herbert B. 2001. *A mathematical introduction to logic.* Second edn. Harcourt/Academic Press, Burlington, MA.

Foreman, Matthew, and Kanamori, Akihiro (eds). 2010. *Handbook of set theory.* New York: Springer-Verlag. In three volumes.

Goldstern, Martin, and Judah, Haim. 1998. *The incompleteness phenomenon.* Natick, MA: A. K. Peters Ltd. Reprint of the 1995 original.

Hrbacek, Karel, and Jech, Thomas. 1999. *Introduction to set theory.* Third edn. Monographs and Textbooks in Pure and Applied Mathematics, vol. 220. New York: Marcel Dekker Inc.

Jech, Thomas. 2003. *Set theory.* Springer Monographs in Mathematics. Berlin: Springer-Verlag. The third millennium edition, revised and expanded.

Kechris, Alexander S. 1995. *Classical descriptive set theory.* Graduate Texts in Mathematics, vol. 156. New York: Springer-Verlag.

Kunen, Kenneth. 1983. *Set theory.* Studies in Logic and the Foundations of Mathematics, vol. 102. Amsterdam: North-Holland. Reprint of the 1980 original.

Moschovakis, Yiannis N. 2009. *Descriptive set theory.* Second edn. Mathematical Surveys and Monographs, vol. 155. Providence, RI: American Mathematical Society.

Royden, H. L. 1988. *Real analysis.* Third edn. New York: Macmillan.

Rudin, Walter. 1976. *Principles of mathematical analysis.* Third edn. New York: McGraw-Hill. International Series in Pure and Applied Mathematics.

Index